Second Edition

DELTAS

PROCESSES OF DEPOSITION & MODELS FOR EXPLORATION

James M. Coleman

Coastal Studies Institute
Louisiana State University

International Human Resources Development Corporation

Boston

Acknowledgments

The delta studies conducted by the Coastal Studies Institute, Louisiana State University, during the period 1953-1975 were supported by the Geography Programs, Office of Naval Research, Arlington, Virginia 22217. Support for a few of the specific studies in the Mississippi River was provided by various petroleum companies and U.S. Geological Survey, Marine Geology Branch, Corpus Christi, Texas. The writer has benefited substantially during the delta studies by association with various colleagues. I especially wish to acknowledge the following colleagues: the late Dr. R. J. Russell, Dr. J. P. Morgan, Dr. W. G. McIntire, Dr. Sherwood Gagliano, Dr. L. D. Wright and Dr. J. C. Ferm. It was through discussions and field work with these colleagues that many of the concepts presented in the manuscript were conceived. Colleagues at the Coastal Studies Institute have freely contributed their time to lively discussions that have often resulted in the identification of significant problems. The staff at CSI, Mrs. Shirley Gerald (editor), Mrs. Gerry Dunn (draftsman), and others, have aided the writer in compilation of the text and illustrations. Dr. George deVries Klein reviewed the monograph.

Copyright © 1982 by International Human Resources Development Corporation. Original Copyright © 1981, 1976 by Burgess Publishing Company. All rights reserved. No part of this book may be used or reproduced in any manner whatsoever without written permission of the publisher except in the case of brief quotations embodied in critical articles and reviews. For information address: IHRDC, Publishers, 137 Newbury Street, Boston, MA 02116.

ISBN: 0-934634-26-2 Cloth
ISBN: 0-934634-27-0 Paper

Library of Congress Catalog Card Number: 80-70004

Printed in the United States of America

Cover Photo: ERTS satellite imagery of Mississippi River Delta 1972 flood.

CONTENTS

1. **Introduction,** 1
2. **Deltaic Processes,** 10
    1. Climate, 10
    2. Water Discharge, 12
    3. Sediment Yield, 12
    4. River Mouth Processes, 13
    5. Waves and Their Effects, 17
    6. Tidal Processes, 20
    7. Wind System, 21
    8. Littoral Currents, 22
    9. Shelf Slope, 23
    10. Tectonics of the Receiving Basin, 24
    11. Receiving Basin Geometry, 24
    12. Summary, 26
3. **Mississippi River Delta,** 28
    1. Introduction, 28
    2. Environments of Deposition, 31
    3. Lateral and Vertical Relationships of Deltaic Facies, 53
    4. Subaqueous Deformational Processes: Mississippi River Delta Plain, 56
4. **Variability of Modern Deltas,** 85
    1. Mississippi River Delta, 88
    2. Klang River Delta, 90
    3. Ord River Delta, 95
    4. Burdekin River Delta, 99
    5. Sao Francisco River Delta, 103
    6. Senegal River Delta, 106
    7. Summary, 107

References Cited, 112

Series Editor: George deVries Klein
Department of Geology
University of Illinois at Urbana-Champaign
Urbana, IL 61801

Consulting Editors:

George Brimhall
Department of Geology and Geophysics
University of California—Berkeley
Berkeley, CA 94720

William J. Hinze
Department of Geosciences
Purdue University
West Lafayette, IN 47907

W. Stuart McKerrow
Department of Geology and Mineralogy
University of Oxford
Parks Road
Oxford OX1 3PR United Kingdom

J. Casey Moore
Earth Sciences
University of California—Santa Cruz
Santa Cruz, CA 95064

1 INTRODUCTION

Deltaic depositional facies result from interacting dynamic processes (wave energy, tidal action, climate, etc.) which modify and disperse riverborne clastics. Since ancient times, river deltas have been of fundamental importance to civilization. Owing to their early significance as agricultural lands, deltas received considerable attention from scholars such as Homer, Herodotus, Plato, and Aristotle. In more recent times, subsurface deltaic facies have played a paramount role in accommodating the world's energy needs; ancient deltaic sediments have provided source beds and reservoirs for a large percentage of the known petroleum reserves. The facies relationships, sand body distribution, and mechanisms responsible for development of deltaic sand bodies must be understood before efficient exploration of these sand bodies can be accomplished.

The term delta was first applied by the Greek historian Herodotus, approximately 450 B.C., to the triangular alluvial deposits at the mouth of the Nile River. In broader terms, deltas can be defined as those coastal deposits, both subaqueous and subaerial, derived from riverborne sediments. Also included are those sedimentary deposits molded by various marine agents such as waves, currents, and tides that are found within the deltaic plain. Because the different processes which control delta development vary appreciably in relative intensity on a global scale, delta plain landforms span nearly the entire spectrum of coastal features and include distributary channels, river mouth bars, interdistributary bays, tidal flats, tidal ridges, beaches, dunes, dune fields, swamps, marshes, and evaporite flats.

Modern-day deltas exist in a wide variety of settings. Some deltas occur along coasts which experience negligible tides and minimal wave energy, whereas others are formed in the presence of extreme tide ranges or are formed against the opposing forces of intensive wave action. Deltas may accumulate in the humid tropics where vegetation is abundant and biological and chemical processes are of prime importance or they may form in arid or arctic environments where biologic activity is subdued. Despite the various environmental contrasts, all actively prograding deltas have at least one common attribute: a river supplies clastic sediment to the coast and inner shelf more rapidly than it can be removed by marine processes.

Deltaic deposits are found where a stream debouches into a receiving basin, whether the receiving basin is an ocean, inland sea, bay, estuary, or lake. Various sizes and a variety of plan views can be found in deltas throughout the world. Figure 1.1 and Table 1.1 show locations and characteristics of some of the world's largest modern deltas. Although major deltas occur in all latitudes except at the poles and on the shores of virtually all seas, certain requirements are necessary for their occurrence.

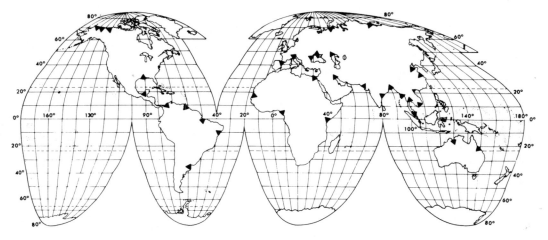

Figure 1.1. Location of major modern deltas. (Republished by permission of the Houston Geological Society; Coleman and Wright, 1975.)

Table 1.1. Delta Characteristics

	Amazon (Brazil)	Burdekin (Australia)	Chao Phraya (Thailand)	Colville (Alaska)	Danube (Romania)	Dneiper (USSR)
DRAINAGE BASIN						
Basin Climate	Humid Trop.	Dry Trop.	Humid Trop.	Arctic	Cool Temp.	Cool Temp.
Rainfall (mm)	1,701	662	1,317	116	792	489
Basin Area (x 10^3 sq km)	5,877.5	266.7	92.2	59.5	712.6	801.3
Average Relief (m)	2,050	339	177	469	292	45
Hypsometric Integral	0.083	0.225	0.197	0.324	0.214	—
ALLUVIAL VALLEY						
Channel Length (km)	—	99	—	481	774	289
Channel Type	B	S	M	M	M	M
Sinuosity Index	—	1.2	—	8.2	7.8	1.3
DELTAIC PLAIN						
Climate	Humid Trop.	Dry Trop.	Humid Trop.	Arctic	Humid Subtrop.	Dry Steppe
Tectonic Setting	S	S	A	S	S	A
Delta Area (sq km)	467,078	2,112	11,329	1,687	2,740	—
Ratio Subaerial/ Subaqueous Area	6.4	2.1	7.4	0.5	8.6	—
Shoreline Length/ Delta Width	1.20	1.66	1.15	1.73	1.65	1.82
Discharge (m^3/sec)	149,736	475.7	831.4	491.7	6,250.1	1,370.0
Discharge Peakedness Index	1.06	2.43	2.32	1.00	1.50	2.30
Delta Channel Pattern	III	I	III	I	III	I
Ratio Channel Bifurcations/Rejoinings	0.58	0.57	0	0.25	0.46	0.50
Type Delta Switching	—	I	II	II	III	II
Sand Body Type	II	III	II	IV	III	I
RECEIVING BASIN						
Basin Geometry	II	—	II	II	V	II
Offshore Slope (%)	0.5	9.2	0.6	2.1	14.1	4.4
River-mouth Type	Funnel	Straight	Funnel	Straight	Straight	Straight
No. River Mouths	6	4	1	19	14	12
Bar Type	Linear Ridge	Radial	Radial	—	—	—
Av. Wave Power (x 10^7 ergs/sec)	0.193	6.414	0.736	0.001	0.034	—
Tide Range (m)	4.9 SD	2.2 SD	2.4 M	0.21 D	0	0
Offshore Wind — (%)	4.7	22.2	26.2	26.7	45.7	34.7
Alongshore Wind + (%)	0.1	1.3	11.9	0.7	12.9	0
Alongshore Wind — (%)	42.8	38.9	26.5	34.8	1.0	43.9
Onshore Wind (%)	52.5	37.6	38.3	36.9	38.7	21.4

Prerequisite for a significant deltaic accumulation is the existence of a major river system which carries substantial quantities of clastic sediment. Such large river systems require a large drainage basin in which precipitation is abundant. Sediments are supplied by erosion and individual tributaries merge to create larger trunk streams. The sediment water discharge is then transported down the alluvial valley to the coast. Thus the presence of a large delta system depends largely on the nature of the drainage basin itself. Drainage basin climate, geology, relief, and area are all critical characteristics that determine river discharge. Of all the requisites for the occurrence of a major river system, probably the most restrictive is the necessity for a large catchment area. The mean basin area of the rivers listed in Table 1 is approximately 10^5 km^1; the drainage basin of the Mississippi, for example, covers 41% of the continental United States or an area of 3.3×10^6 km^2. The Amazon River system of Brazil arises from a basin 5.9×10^6 km^2 and represents one of the largest drainage basins of modern-day river systems. Therefore, large river systems and consequently large deltas normally are not present along coasts which are highly active tectonically or which are in very close proximity to drainage divides. Accordingly the major river systems are also closely dependent upon global tectonics. Inman and Nordstrom (1971) have produced a macro-scale classification of the world's coasts in which they distinguish between the following tectonic classes and subclasses:

Table 1.1. Delta Characteristics (Continued)

	Ebro (Spain)	Ganges-Brahmaputra (Bangladesh)	Godavari (India)	Grijalva (Mexico)	Hwang Ho (China)	Indus (W. Pakistan)
DRAINAGE BASIN						
Basin Climate	Dry Trop.	Warm Temp.	Dry Trop.	Dry Subtrop.	Dry Subtrop.	Dry Subtrop.
Rainfall (mm)	583	1,814	1,512	1,585	431	720
Basin Area (x 10^3 sq km)	89.8	1,597.2	305.3	112.0	865.1	887.7
Average Relief (m)	402	682	191	351	514	606
Hypsometric Integral	0.317	0.228	0.362	0.221	0.387	0.276
ALLUVIAL VALLEY						
Channel Length (km)	67	952	—	—	609	698
Channel Type	M	B	B	M	B-M	B
Sinuosity Index	1.4	1.2	—	—	1.1	1.5
DELTAIC PLAIN						
Climate	Dry Trop.	Humid Trop.	Dry Trop.	Humid Trop.	Dry Subtrop.	Subtrop. Arid
Tectonic Setting	S	A	S	S	S	A
Delta Area (sq km)	624	105,641	6,322	17,028	36,272	29,524
Ratio Subaerial/Subaqueous Area	4.6	2.4	4.5	6.7	3.3	8.1
Shoreline Length/Delta Width	2.89	1.51	1.76	1.26	1.36	1.42
Discharge (m^3/sec)	552.0	34,500.0	3,180.0	—	2,571.0	4,274.0
Discharge Peakedness Index	1.05	1.66	2.04	1.40	2.04	1.25
Delta Channel Pattern	III	II	I	III	III	I
Ratio Channel Bifurcations/Rejoinings	0	0.36	0.50	0	0	0.40
Type Delta Switching	II	II	II-III	II	II	II
Sand Body Type	III	II	V	V	—	II
RECEIVING BASIN						
Basin Geometry	V	II	IV	IV	I-II	II
Offshore Slope (%)	36.0	1.5	12.8	7.4	1.5	9.6
River-mouth Type	Straight	Straight	Straight	Funnel	Straight	Funnel
No. River Mouths	2	8	11	2	5	4
Bar Type	—	Linear Ridges	—	—	—	—
Av. Wave Power (x 10^7 ergs/sec)	0.155	0.586	—	—	0.218	14.15
Tide Range (m)	0	3.6 SD	1.2 SD	0.79 M	1.13 SD	2.6 SD
Offshore Wind — (%)	59.5	20.4	18.7	38.4	19.6	41.2
Alongshore Wind + (%)	10.4	20.5	21.7	0.0	22.9	0.0
Alongshore Wind — (%)	4.8	33.3	41.7	39.8	7.4	28.8
Onshore Wind (%)	24.4	25.8	18.0	21.8	50.1	30.1

1. Collision Coasts:
   a. Continental collision coasts - collision coasts involving the margin of continents where a thick plate collides with a thin plate (i.e. west coast of the Americas);
   b. Island arc collision coast - collision coast along island arcs where thin plate collides with another thin plate (i.e. the Philippines, the Indonesian and Aleutian Island arcs).
2. Trailing edge coasts:
   a. Neo-trailing edge coasts - new trailing edge coasts formed near beginning separation centers and rifts (i.e. Red Sea, Gulf of California);
   b. Afro-trailing edge coasts - opposite coast of the continent is also trailing so that the potential for terrestrial erosion and deposition at the coast is low (i.e. Atlantic and Indian coast of Africa);
   c. Amero-trailing edge coasts - the trailing edge of a continent with a collision coast that therefore is actively modified by the depositional products and erosional effects from an extensive area of high interior mountains (i.e. east coast of the Americas).
3. Marginal sea coasts - coasts fronting on marginal seas and protected from the open ocean by island arcs (i.e. Vietnam, Southern China, Korea).

Table 1.1. Delta Characteristics (Continued)

	Irrawaddy (Burma)	Klang (Malaysia)	Lena (USSR)	Mackenzie (Canada)	Magdalena (Columbia)	Mekong (Vietnam)
DRAINAGE BASIN						
Basin Climate	Humid Subtrop.	Humid Trop.	Humid Arctic	H. Subarctic	Humid Trop.	Humid Trop.
Rainfall (mm)	2,192	2,135	253	257	1,777	1,530
Basin Area (x 10^3 sq km)	341.8	0.9	2,421.4	1,448.4	251.7	517.5
Average Relief (m)	353	446	223	730	787	440
Hypsometric Integral	0.163	0.146	0.309	0.321	0.255	0.213
ALLUVIAL VALLEY						
Channel Length (km)	208	—	—	27	136	325
Channel Type	B	M	B	S	M	M-B
Sinuosity Index	1.5	—	—	1.1	1.1	1.3
DELTAIC PLAIN						
Climate	Humid Trop.	Humid Trop.	Arctic	Arctic	Humid Trop.	Humid Trop.
Tectonic Setting	A	S	A	A	A	A
Delta Area (sq km)	20,571	1,817	43,563	8,506	1,689	93,781
Ratio Subaerial/ Subaqueous Area	0.1	0.8	1.2	1.7	14.8	2.0
Shoreline Length/ Delta Width	1.65	1.35	1.85	—	1.02	2.20
Discharge (m³/sec)	12,558	1,100	1,402	8,583	7,500	14,168
Discharge Peakedness Index	1.61	—	3.56	1.35	—	1.55
Delta Channel Pattern	II	II	II	II	III	III
Ratio Channel Bifurcations/Rejoinings	0.83	0	0.40	0.36	0	0.86
Type Delta Switching	III	II	—	II	II	III
Sand Body Type	III	II	—	I	V	III
RECEIVING BASIN						
Basin Geometry	II	I	—	II or IV	V	IV
Offshore Slope (%)	1.4	4.1	0.8	1.9	36.9	4.3
River-mouth Type	Funnel	Straight	Funnel	Straight	Straight	Funnel
No. River Mouths	10	2	7	11	1	5
Bar Type	—	—	—	—	—	—
Av. Wave Power (x 10^7 ergs/sec)	0.193	0.218	—	—	206.25	—
Tide Range (m)	2.7 SD	4.2 SD	0.21 SD	0.34 D	1.1 D	2.6 M
Offshore Wind — (%)	22.8	12.4	35.6	24.8	4.6	19.5
Alongshore Wind + (%)	16.2	12.0	9.9	23.6	0.0	35.3
Alongshore Wind — (%)	32.3	23.7	13.8	32.8	57.4	32.8
Onshore Wind (%)	28.8	38.9	37.9	19.6	38.3	23.8

Introduction 5

In connection with this classification some 58 major rivers having drainage areas in excess of 10^5 km^2 exist; of these approximately 47% were found to debouch along Amero-trailing edge coasts, 35% along marginal sea coasts, 9% along Afro-trailing edge coasts, and 2% along Neo-trailing edge coasts. Along collision coasts where tectonic activity is high and drainage divides are characteristically close to the sea, only 7% of the rivers examined exist under these conditions. Thus distribution of the land masses in ancient geologic times has had significant control on the existence of modern-day deltas as well as the existence of ancient deltaic deposits.

A river system consists of four primary components: drainage basin, alluvial valley, deltaic plain, and receiving basin (Figure 1.2). The drainage basin basically is a source of the water and sediments, and processes within this component determine the sediment water supply and the initial size and composition of the sedimentary load. The tributaries of the drainage basin eventually merge into one or more major channels and the alluvial valley is formed. The alluvial valley is essentially a conduit in which the river flows over and through its own deposits. In large river systems over long periods of time, sediments have been accumulating within the alluvial valley. As a result, shifting of sediments within the valley is often a significant process which alters the size and composition of the sediment suite entering the valley from the drainage basin. At some point along its length the river ceases to function primarily as a transporting agent and becomes a dispersal system. The

Table 1.1. Delta Characteristics (Continued)

	Mississippi (USA)	Niger (Nigeria)	Nile (Egypt)	Ord (Australia)	Orinoco (Venezuela)	Parana (Brazil)
DRAINAGE BASIN						
Basin Climate	Temperate	Humid Trop.	Dry Subtrop.	Desert	Humid Trop.	Humid Subtrop.
Rainfall (mm)	1,018	1,062	870	528	1,434	1,205
Basin Area (x 10^5 sq km)	3,344.6	1,112.7	2,715.6	78.0	951.3	2,871.8
Average Relief (m)	915	93	1,007	297	243	1,110
Hypsometric Integral	0.241	0.212	0.168	0.662	0.095	0.100
ALLUVIAL VALLEY						
Channel Length (km)	—	166	2,577	69	520	—
Channel Type	M	B	M	M	M	B
Sinuosity Index	—	1.0	1.6	1.5	1.2	—
DELTAIC PLAIN						
Climate	Humid Subtrop.	Humid Trop.	Arid Desert	Dry Subtrop.	Humid Trop.	Humid Subtrop.
Tectonic Setting	A	A	I	S	I	I
Delta Area (sq km)	28,568	19,135	12,512	3,896	20,642	5,440
Ratio Subaerial/Subaqueous Area	5.3	8.5	9.0	3.2	8.6	1.5
Shoreline Length/Delta Width	2.03	1.24	1.20	2.28	1.80	1.81
Discharge (m^3/sec)	15,631	8,769	1,480	166	25,200	12,658
Discharge Peakedness Index	0.75	2.38	2.65	2.43	1.12	21.7
Delta Channel Pattern	I	II	I	III	II	I
Ratio Channel Bifurcations/Rejoinings	0.26	0.84	0.96	0	0.29	0.29
Type Delta Switching	I	II-III	II	II	I	I
Sand Body Type	I	III	III-IV	II	III	I
RECEIVING BASIN						
Basin Geometry	V	III-IV	I-II	II	I-II	II
Offshore Slope (%)	7.0	6.2	7.3	3.9	2.8	0.3
River-mouth Type	Straight	Constricted	Straight	Funnel	Funnel	Straight
No. River Mouths	22	11	2	2	18	21
Bar Type	—	Radial & Jet.	—	—	—	—
Av. Wave Power (x 10^7 ergs/sec)	0.034	2.007	10.25	1.062	—	—
Tide Range (m)	0.43 M	1.4 D	0.43 SD	5.8 SD	1.8 SD	0.64 SD
Offshore Wind — (%)	26.0	2.92	15.7	26.7	55.0	46.5
Alongshore Wind + (%)	28.2	16.2	33.6	17.2	42.3	12.6
Alongshore Wind — (%)	0.3	1.2	0.0	42.0	0.0	12.8
Onshore Wind (%)	44.7	79.9	50.8	14.2	2.8	36.9

sediments accumulate and form the deltaic plain. This component of a river system results from the interaction between riverine and marine processes. The morphology and geometry of the delta reflect: the hydraulic regime; sediment load; geologic structure; and tectonic stability; climate; tides; winds; waves; water density contrasts; and coastal currents; and the innumerable interactions of all of these factors. The receiving basin is the recipient of the riverborne clastic deposits. The characteristics of the receiving basin are as important to the development of a delta as is the river, and the various oceanic processes active near the delta coast are crucial to the shaping and remolding of the riverborne deposits.

The delta plain itself can be subdivided into basic physiographic zones as illustrated in Figure 1.3. Every delta consists of a subaqueous and subaerial component even though the relative areas of these may vary considerably. The subaerial delta is that portion of the delta plain above the low tide limit. It experiences a broad range of processes and displays a complex assemblage of depositional forms and environments. In many cases the deposits may be little more than a thin veneer capping a much thicker sequence of subaqeous deltaic sequences. The upper portion of the subaerial delta, commonly referred to as the upper deltaic plain, is normally the older portion of the subaerial delta and exists above significant tidal or marine influence. It is essentially the seaward continuation of the alluvial valley and is dominated by riverine depositional processes. The lower portion of the deltaic plain lies within the realm of river-marine interaction and extends landward from the low-tide

Table 1.1. Delta Characteristics (Continued)

	Pechora (USSR)	Po (Italy)	Red (N. Vietnam)	Sagavanirktok (Alaska)	Sao Francisco (Brazil)	Senegal (Senegal)
DRAINAGE BASIN						
Basin Climate	Subarctic	Temperate	Humid Subtrop.	Arctic	Humid Trop.	Dry Subtrop.
Rainfall (mm)	446	847	1,282	116	1,223	1,381
Basin Area (x 10^2 sq km)	300.7	71.7	143.9	11.8	602.3	196.4
Average Relief (m)	147	480	420	324	63	40
Hypsometric Integral	0.171	0.258	0.571	0.404	0.452	0.431
ALLUVIAL VALLEY						
Channel Length (km)	254	107	57	55	150	648
Channel Type	B	M	S	B	B	M
Sinuosity Index	1.2	2.1	1.2	1.1	1.1	1.5
DELTAIC PLAIN						
Climate	Subarctic	Temperate	Subtrop.	Arctic	Dry Trop.	Arid Desert
Tectonic Setting	—	A	S	S	S	S
Delta Area (sq km)	—	13,398	11,908	1,178	734	4,254
Ratio Subaerial/						
Subaqueous Area	—	8.0	3.9	0.9	1.4	20.1
Shoreline Length/						
Delta Width	1.46	1.39	1.33	1.64	1.08	1.02
Discharge (m³/sec)	3,362	1,484	3,913	—	3,420	867.8
Discharge Peakedness Index	—	1.18	1.70	—	1.48	2.30
Delta Channel Pattern	I-II	I	III	II	III	III
Ratio Channel						
Bifurcations/Rejoinings	0.667	0	0.58	0.17	0	1
Type Delta Switching	—	—	III	II	II	II
Sand Body Type	—	—	III	IV	V	VI
RECEIVING BASIN						
Basin Geometry	II	II	V	—	IV	IV
Offshore Slope (%)	6.2	5.6	3.7	3.5	11.2	17.0
River-mouth Type	Straight	Straight	Funnel	Straight	Constricted	Reflected
No. River Mouths	10	8	12	11	1	1
Bar Type	—	—	—	—	Radial	—
Av. Wave Power						
(x 10^7 ergs/sec)	—	—	—	—	30.415	112.42
Tide Range (m)	0.73 SD	0.73 SD	1.9 D	0.21 M	1.86 SD	1.22 SD
Offshore Wind — (%)	13.4	44.8	15.4	25.3	2.9	32.4
Alongshore Wind + (%)	5.8	23.4	28.9	4.9	49.6	0.0
Alongshore Wind — (%)	40.5	9.6	26.3	24.3	0.0	65.3
Onshore Wind (%)	40.3	20.5	30.0	45.7	47.5	3.1

Introduction

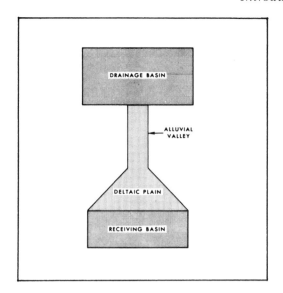

Figure 1.2. Components of a river system. (Republished by permission of the Houston Geological Society; Coleman and Wright, 1975.)

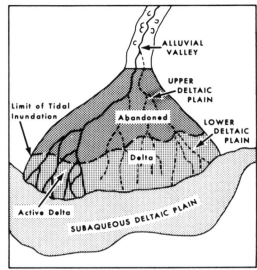

Figure 1.3. Components of a delta plain.

Table 1.1. Delta Characteristics (Continued)

	Shatt-al-Arab (Iraq)	Tana (Kenya)	Volga (USSR)	Yangtze-Kiang (China)
DRAINAGE BASIN				
Basin Climate	Dry Subtrop.	Subtrop.	Dry Subtrop.	Dry Subtrop.
Rainfall (mm)	148	733	527	1,215
Basin Area (x 10^3 sq km)	461.7	63.5	1,614.4	1,354.2
Average Relief (m)	112	555	32	567
Hypsometric Integral	0.260	0.143	—	0.397
ALLUVIAL VALLEY				
Channel Length (km)	—	—	381	266
Channel Type	M	M	M	B
Sinuosity Index	—	—	1.2	1.1
DELTAIC PLAIN				
Climate	Dry Trop.	Dry Subtrop.	Dry Subtrop.	Humid Subtrop.
Tectonic Setting	A	S	I	I-S
Delta Area (sq km)	18,497	3,659	27,224	66,669
Ratio Subaerial/				
Subaqueous Area	0.6	1.0	2.0	1.7
Shoreline Length/				
Delta Width	1.30	1.12	1.93	1.55
Discharge (m^3/sec)	1,300	172	7,736	—
Discharge Peakedness Index	2.66	1.67	3.10	—
Delta Channel Pattern	III	I	I-II	III
Ratio Channel				
Bifurcations/Rejoinings	0	0	0.77	0.83
Type Delta Switching	II-III	I	I	—
Sand Body Type	II	V	I	—
RECEIVING BASIN				
Basin Geometry	II	IV	V	III
Offshore Slope (%)	0.470	0.032	—	0.013
River-mouth Type	Funnel	Deflected	Straight	Straight
No. River Mouths	2	1	15	3
Bar Type	—	—	—	—
Av. Wave Power				
(x 10^7 ergs/sec)	—	—	—	0.127
Tide Range (m)	2.5 SD	2.9 SD	0	3.7 SD
Offshore Wind — (%)	35.8	21.3	—	48.2
Alongshore Wind + (%)	25.9	15.3	—	7.5
Alongshore Wind — (%)	30.7	14.0	—	24.2
Onshore Wind (%)	7.6	49.4	—	20.2

mark to the limit of tidal influence. The lower plain is most extensive where tidal range is large and seaward gradients of the river channel and delta are low. The subaqeous delta is that portion of the delta plain which lies below the low-tide water level. It is the foundation on which progradation of the subaerial delta must proceed. Most commonly the subaqeous delta is characterized by a seaward fining of sediments, sand, and coarser clastics being deposited near the river mouth and finer-grained sediments settling further offshore from suspension in the water column. The seaward-most portion of the subaqeous delta is normally referred to as prodelta deposits and is composed of the finest material deposited from suspension. The prodelta clays grade landward and upward into the coarser silts and sands of the delta front. Immediately at the mouths of the active distributaries are found the coarsest sand deposits, commonly referred to as the distributary mouth bar deposits.

The delta plain can also be subdivided into the active and abandoned zones shown in Figure 1.3. The active delta plain is the actively accreting portion occupied by the functioning distributary channels. The abandoned delta plain results when the river changes its lower course to a shorter or more efficient route to the sea, a process which causes a corresponding shift in the locus of river mouth sedimentation. The abandoned portion of the delta then becomes subjected to reworking by marine processes. In some cases, if wave reworking is pronounced, the delta shoreline will undergo a landward transgression and the result will usually take the form of a coastal barrier, beach, or dune complex. In other instances, when subsidence of the delta deposits is high, the sea will encroach rapidly across the abandoned delta surface, resulting in the deposition of shallow marine deposits directly on top of the former delta surface without significant reworking of the regressive sequence.

The processes which occur in the active delta are of utmost importance, for it is these processes which determine the development and distribution of the various facies relationships displayed by modern deltas. In most instances, the active delta consists of one or more river mouth systems which are the dissemination points for sediments which contribute to delta progradation. When a stream discharges into a receiving basin, its momentum is dissipated by the interaction of the river water with the ambient seawater; the result is the deceleration of the affluent, a consequent loss of sediment transporting ability and deposition. Commonly, there is a progressive seaward decrease in the concentration and grain size of sediments transported by the affluent. Figure 1.4, modified from Scruton (1960), illustrates the general process of river mouth sedimentation and the seaward fining of sediments. Most rapid deposition and deposition of the coarsest materials takes place a short distance from the river mouth, a sand body commonly referred to as a distributary mouth bar. Seaward of this region, the percentage of sand diminishes rapidly and a zone of interfingering sands, silts, and clays is commonly found. This zone is commonly referred to as the delta front deposits. Beyond or seaward of the delta front, only fine silts and clays remain in suspension and are deposited out of suspension on the continental shelf. These deposits are commonly referred to as prodelta clays. Thus, as can be seen in Figure 1.4, there is a gradation in sediment from sand deposits near the river mouth to finer grain sequences in a seaward direction. With time, and as the delta builds or progrades seaward, the coarser sand deposits are progressively laid down over the finer-grained deposits, resulting in the classical coarsening-upward sequence common to all delta deposits.

If the sediments deposited seaward of the river mouth accumulate faster than subsidence, or removal of sediments by marine processes, deltaic progradation will take place. As the delta system progrades, subaerial deltaic deposits will form atop the uppermost portions of the subaqueous delta deposits, forming a complete delta sequence. Those deposits marginal to, and between the distributary channels, comprise the largest percentage of the total subaerial delta. Immediately adjacent to the active channels, the most common depositional forms are probably natural levees, overbank splays, and associated minor channel systems. Further from the channel in the interdistributary

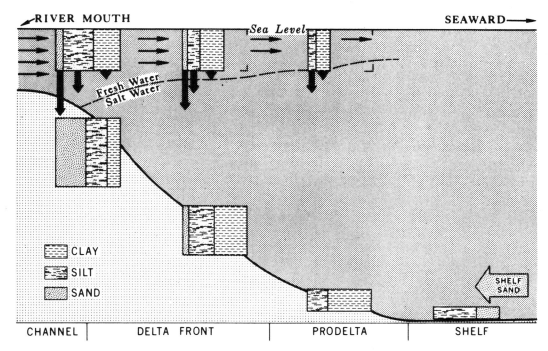

Figure 1.4. Deposition of sediment at a river mouth. (Modified from Scruton, 1960.)

regions, a wide variety of depositional environments can be found. Rapid distributary progradation coupled with rapid subsidence and low wave energy results frequently in the formation of shallow-water interdistributary bays. If subsidence is slow and depositional rates high, the interdistributary areas may be completely infilled with marsh or mangrove deposits. High tide range and arid climate often yield interdistributary evaporite or barren salt flats where intricate networks of tidal creeks separated by broad evaporitic sequences occur. In some instances, in regions experiencing extremely high wave action, continuous sandy plains of successively closely spaced beach ridges and dunes will occupy the interdistributary region.

Thus modern deltas all display common components, but their size, shape, and subsurface relationships vary from delta to delta. In order to interpret and exploit efficiently ancient deltaic sequences, it is of fundamental importance to understand the mechanisms responsible for controlling the formation and distribution of deltaic sand bodies. This aspect, then, forms a major and prime purpose of the following text.

2 | DELTAIC PROCESSES

Interactive coastal processes active during the time of deposition exert significant control over the distribution, orientation, and internal geometry of deltaic sand bodies. Some of the major process controls of a river system are illustrated in Figure 2.1. Coleman and Wright (1971, 1975) discuss the various aspects of these dynamic interacting coastal processes and their effect and significance in controlling delta formation. Although all processes influence to some degree the relationships of various subenvironments within deltaic sequences, only a relatively small number of these processes strongly influence the formation of various types of delta sand bodies. The most important of the processes are climate, water and sediment discharge and its variability, sediment yield, river mouth processes, nearshore wave power, tides and tidal regime, winds, nearshore currents, shelf slope, tectonics of the receiving basin, and receiving basin geometry.

1. Climate

Climate more than any other single factor determines the variation and intensity of physical, chemical, and biological processes active within all components of the river system. Runoff is the function of precipitation and the volume of water available for runoff is dependent upon the difference between actual precipitation and the potential evapo-transpiration. In large tropical basins precipitation is normally high in relationship to evapo-transpiration and the result is the continual supply of sediments and water to the alluvial valley. Chemical weathering and mechanical breakdown of rock proceed rapidly and result in high sediment yields. In general, those basins characterized by tropical climate contain high suspended sediment loads per equal area of basin. The volume of sediment water mixture delivered to the alluvial valley is not extremely erratic and the channels normally display rather stable patterns. In arid or arctic drainage basins precipitation is normally erratic and vegetation is sparse. Evaporation is normally high most of the year and sediment-water yield to the alluvial valley is sporadic. Channels in such settings tend to migrate considerably and be extremely unstable. Braided channel patterns are the most common and in most instances, the ratio of bedload to suspended load is quite high.

One of the major roles played by climate within the delta plain and the alluvial valley is in controlling the composition and quantity of in situ deposits. Sand bodies that form within the delta plain are normally encased in finer-grained deposits and the nature of the subaerial deposits that cap sand bodies is highly dependent upon climatic conditions.

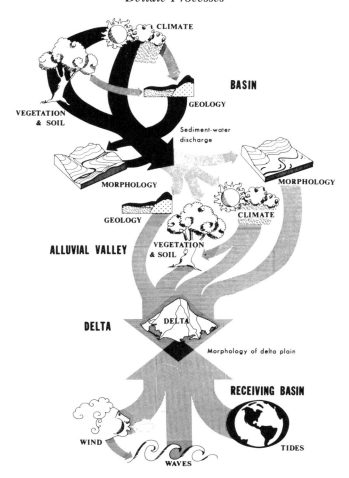

Figure 2.1. Major process controls on a river system. (Republished by permission of the Houston Geological Society; Coleman and Wright, 1975.)

Tropical and arctic environments are conducive to the production and preservation of organic material and commonly peat deposits form the bulk of the subaerial deposits. High biological bioturbation is common in all of the subaerial environments. The abundance of organic detritus combined with a favorable environment for preservation results in accumulation of thick transported organic detritus within the sand bodies. Many of the large transported coal seams in ancient deltaic sand bodies accumulated under tropical or arctic conditions. Temperate climates also tend to produce peat, but organic production and preservation is generally less than in the tropical areas; there, peat deposits tend to be thin, display wide lateral continuity, and quite often interfinger strongly with fine-grained clastic deposits. The rapid degradation of in situ organic deposits releases complex chemical solutions to the water column and to pore waters. As a result, diagenetic products (cementation, nodule formation, etc.) occur rapidly and in many instances prevent excessive compaction of the delta sequences. In arid conditions evaporation begins to play a significant role. Evaporites and chemical precipitants begin to form the major deposits that cap delta sand bodies. Under present-day conditions, numerous modern deltas display a high percentage of evaporites in the total volume of the subaerial deltaic plain deposits and thus under these conditions, evaporitic sequences play an important role in this type of delta setting.

2. Water Discharge

The river discharge regime depends on the climatic factors active within the drainage basin. Temporal discharge tendencies and variation in discharge throughout a hydrologic year often exert a great influence on alluvial valley and deltaic sand body geometries. When variations in annual discharge are small, the channel will be better able to adjust to an equilibrium configuration and in these instances, stable channels, primarily of the meandering type, will result. Alluvial valley sand bodies tend to take the form of shoestring-type sand bodies. When water discharge is highly variable and extremely erratic, channels do not have sufficient time to adjust to any given flow and as a result may be unstable much of the year. Braided patterns often result from this situation and the river tends to change its course frequently and migrate rapidly and erratically. This type of channel migration often produces sheet sands associated with broad, constantly shifting braided channels. Sand bodies tend to be thin, on the order of 10 to 20 m thick, but cover considerable areal extent.

The flow distribution also tends to affect the size and sorting of the sediment load. Extremely erratic discharge in rivers most often results in relatively coarse and poorly sorted sediments. Sand bodies are often clay-bound and show little or no permeability. Less erratic discharge normally results in a tendency for the sand bodies to display better sorting. Permeability is generally good, and sand bodies have regular patterns. Meander-belt bodies, produced by regular discharge patterns, tend to show a sand sequence which consistently grades from coarse to fine upwards; whereas braided channel sands produced by more erratic discharge display considerable grain-size variation both laterally and within the vertical sequence. Such sand bodies are characterized by multiple fining-upward sequences which are stacked one upon another and tend to display no regular pattern. Quite often permeability and porosity can vary laterally significantly over distances of hundreds of meters.

In the delta plain, discharge tendencies are especially important to the rate and pattern of delta growth. Sand body distribution within the delta depends upon the rate of supply of sediment to the coast by the river and the ability of wave and current forces to rework and redistribute the sediment. In those deltas where sediment and water discharge are constant and high year-long (Mississippi), long linear sand bodies are common, trending at high angles to the coast. In other cases where sediment yield and water discharge variations are extreme throughout the year, sands delivered to the coast have ample opportunity to be reworked by marine processes and often linear sand bodies tend to form parallel to the shoreline trend.

3. Sediment Yield

Factors controlling sediment yield are numerous and complex, but primarily are a function of the basin area and discharge. Rivers characterized by high fine-grained sediment loads tend to build extensive subaqueous delta platforms which are composed of fine-grained, high water content, unstable clays. These deltas are normally characterized by a wide variety of deformational features such as slumping, clay flowage, and local diapirism. Compaction rates are normally high and in many instances, sand bodies are deformed by compactional type structures. Those river systems characterized by high bedload sediment yields often display complex, interfingering, scoured-base sand bodies throughout the alluvial valley and deltaic plain. Often a high percentage of the total deltaic plain sediments consists of reworked channel deposits. Rivers such as the Nile, the Niger, and the Burdekin show fining-upward sequences of channel sands well within the delta proper.

4. River Mouth Processes

The river mouth is the point at which the seaward-flowing water leaves the confines of the channel banks and spreads and mixes with ambient waters of the receiving basin. This point is the dynamic dissemination for sediments which contribute to continued delta progradation and is responsible for forming one of the major sand bodies associated with deltaic sequences, the distributary mouth bar deposit. The geometry of the river mouth and the distributary bar topography together make up a single unit which both influences and is influenced by effluent dynamics. The resulting geometry and distribution of river mouth bar sand bodies is determined by riverine flow condition, density contrasts between issuing and ambient water, water depths, bottom slope seaward of the mouth, tidal range and tidal currents within the lower river channel, and the ability of waves and other forces to obstruct the outlet. Comprehensive studies, both theoretical and field, of river mouth processes have been completed by Bates (1953), Borichanski and Mikhailov (1966), Jopling (1963), Scruton (1960) and Wright and Coleman (1974).

The most important river mouth processes are those by which the river effluent interacts with ambient water within the receiving basin, resulting in deconcentration of outflow momentum and consequent loss of transporting ability. This interaction yields specific geometry to the resulting distributary mouth bar deposit. The geometry of the resulting sand body depends upon the relative roles of three primary forces: (1) the inertia of issuing river water and associated turbulent diffusion; (2) friction between the effluent and the bed immediately seaward of the mouth; (3) buoyancy resulting from density contrasts between issuing and ambient fluids. Although all three forces are normally operative to some extent in all river mouths, the relative significance of each depends on outflow velocity, density stratification, and outlet geometry. Figure 2.2 illustrates these three

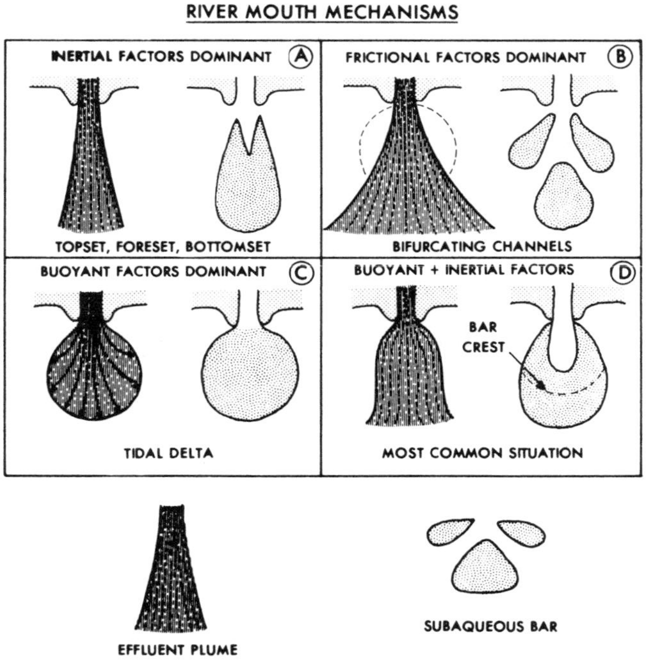

Figure 2.2. River mouth mechanisms.

mechanisms and the resulting plan view geometry of the distributary mouth bar. A fourth case (Figure 2.2), which is common in many modern rivers, indicates interactions that can exist between two of these forces, buoyancy and inertial factors.

When outflow velocities are high, depths immediately seaward of the mouth are relatively large, density contrasts are negligible, inertial forces will be dominant and the effluent will spread and diffuse as a turbulent jet (Figure 2.2a). When this force is the dominant mechanism, distributary mouth bars are generally linear in nature, spread laterally a minimum distance and are generally quite thick. Coarse sediments fall out immediately in the vicinity of river mouths and quite commonly the vertical sequence in such a deposit changes rapidly from fine-grained marine clays upward into a thin silty zone and immediately into coarser river mouth bar sands. Lateral continuity of the sand deposit is normally low. If water depths seaward of the mouth are shallow, turbulent diffusion becomes restricted to the horizontal and bottom friction plays a major role in causing effluent deceleration and expansion (Figure 2.2b). The rapid rate of effluent expansion characteristic of this type of river mouth produces initially a broad arcuate radial bar. However, as deposition on the bar continues, natural subaqueous levees develop beneath the lateral boundaries of the expanding effluent where velocity gradients are steepest. Development of the levees tends to inhibit further increases in effluent expansion rate so that with continuing bar accretion, continuity can no longer be maintained simply by increasing effluent width. As a central portion of the bar grows upward, channelization develops along the threads of maximum turbulence which tend to follow the subaqueous levees. This process results in formation of a bifurcating channel which has a triangular middle ground shoal separating the diverging channel arms as shown in Figure 2.2b. Vertical sequences within this type of distributary mouth bar normally tend to show a coarsening-upward sequence until the top of the distributary mouth bar is reached. Capping this sequence of sands, a finer-grained silt and sandy unit is normally found characterized by the abundance of various types of climbing ripple lamination. These uppermost sediments represent the subaqueous natural levees. This type of distributary mouth bar is extremely common in many of the major modern river deltas.

Most of the world's major rivers debouch into salt-water basins. Normally fresh river water has an average density that approximates 1.00 g/cm^3, whereas seawater has a density of about 1.028 g/cm^3. Although suspended material increases the density of freshwater slightly, the presence of suspended load in most rivers creates a density difference that is small in comparison to that created by the salt content of the marine water. Thus, most river water can be expected to be lighter than the ambient basin water. Strong vertical density gradients exist at the mouths of many rivers and the outflow spreads as a buoyant layer above the underlying salt water (Figure 2.2c). In such instances, buoyancy becomes of paramount importance. Turbulent diffusion tends to be generally suppressed and lateral effluent expansion occurs largely as a result of the buoyant spreading of the freshwater as a relatively homogeneous layer. Deceleration of the outflowing and spreading freshwater effluent decreases radially away from the river mouth exit; coarser sands are deposited in the immediate vicinity of the river mouth and finer-grained sediments are deposited in areas seaward of the river mouth. The river mouth bar deposit then spreads radially away from the river mouth. Normally these distributary mouth bar deposits are relatively thin, but show high lateral continuity. In many instances, the sand sheets will merge from one river mouth to another and the entire front of the delta will be composed primarily of a thin sheet sand formed by coalescing river mouth bar deposits. In all cases cited above, however, the coarsest sediment is deposited near the river mouth and the finer-grained sediments settle further offshore. This situation leads to the formation of a graded base sand body and a coarsening-upward sequence.

Various combinations of these three major forces can exist in modern world deltas. Examination of numerous modern deltas has resulted in establishing five major types of river mouth bars: radial, lunate, middle ground, subaqueous jettied, and tidal ridge and swale bars. These five types of river mouth bars are illustrated in Figure 2.3. Radial bars are common at river mouths where frictional forces are dominant and buoyancy plays a minor role (Figure 2.3a). A bulge in the subaqeous contours is prominent and sands are disseminated several kilometers laterally away from the river mouth. These river mouth bars display extremely well-sorted sands in the upper parts of the sand body and grade rather abruptly offshore into finer-grained silts and clays. The lunate bar (Figure 2.3b) is characterized by a deep scour pool immediately seaward of the river mouth and a shoal or bar developed further seaward. Sands are generally less widespread than in the radial type and commonly the channel will scour through its own sand body into underlying marine prodelta clays. This bar type forms in higher gradient streams where inertial forces play the dominant role. The middle ground bar or bifurcating type (Figure 2.3c) is most commonly found in situations where friction and buoyancy are significant. Sands at the river mouth under these conditions are extremely widespread and the channels that cut through the bar are normally shallow and do not scour completely through the sand body. Subaqueous natural levees are prominent features and the fine-grained nature of these capping deposits makes them excellent seals for the underlying coarser bar deposits. The subaqueous, jettied type of river mouth bar (Figure 2.3d) displays sands that are localized in the immediate vicinity of the river mouth; this type forms where inertial forces are dominant. River mouths are normally deep and quite often are associated with salt water intrusion into the river channels.

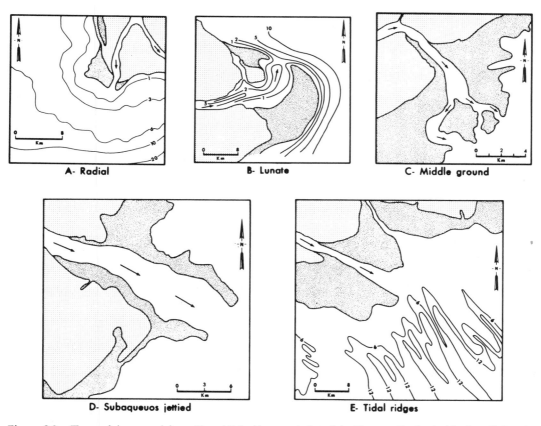

Figure 2.3. Types of river mouth bars. (Republished by permission of the Houston Geological Society; Coleman and Wright, 1975.)

Many river mouths debouch into narrow, elongate basins where high tide range creates strong bidirectional transport. This process leads to the formation of large linear tidal ridges immediately seaward of the river mouth (Figure 2.3e). These ridges are commonly composed of coarse riverborne sand, and vary in size, some displaying heights in excess of 30 m above the adjacent swales. Commonly the ridges are oriented parallel to the river channel. Prominent tidal ridges occur at the mouths of the Ganges-Brahmaputra, the Ord, the Colorado, the Shatt-al-Arab, the Indus, and the Irrawaddi Rivers.

Although the river mouth bar types described above can be found at individual river mouths, many deltas display multiple river mouths and hence have varying types of distributary networks. Three major types of delta channel patterns can be documented, and these are shown diagrammatically in Figure 2.4. The first type (Figure 2.4a) consists of seaward-bifurcating channels, and deltas displaying this channel pattern are characterized by a large number of river mouths. In many instances the river mouths are close enough to one another so that the sands accumulated at the mouth of each distributary channel merge and form a sand sheet. A high subsidence rate, low wave action, low offshore slope, small tidal range, and normally a finer-grained sediment load favor development of this distributary pattern. A second type of distributary pattern (Figure 2.4b) is characterized by bifurcations displaying complex rejoining of the channels and hence fewer active river mouths are present at the shoreline. Conditions that favor development of this distributary pattern include erratic discharges, intermediate wave energy, high tide range, and relatively steep offshore slopes. Modern delta examples displaying this pattern include the Irrawaddi and Nile Rivers. The third type of channel pattern (Figure 2.4c) is relatively simple, consisting of a single channel or relatively few channels all originating from a nearby common point at the head of the delta. In these instances, the river mouths normally display a bell-shape, and lunate bars or tidal ridges are common. High wave action, high tidal ranges, and steep offshore slopes are normally associated with this type of distributary pattern and long linear sand bodies commonly result.

Therefore, the processes operative at the mouths of delta distributaries are critical factors in the initial development of the geometry and distribution of mouth sand bodies,

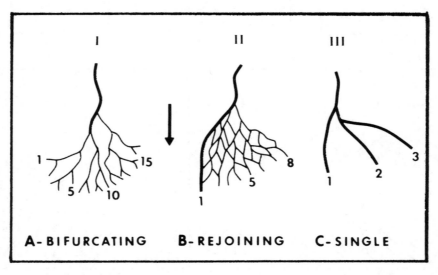

Figure 2.4. Types of distributary channel patterns (Reprinted by permission of the Houston Geological Society; Coleman and Wright, 1975.)

and of the more prominent sand bodies commonly found in most deltas. Much of the hydrocarbon production from deltaic sequences is produced from distributary mouth bar deposits and in outcrops these sand deposits normally form prominent ridges and exposures.

5. Waves and Their Effects

Many of the modern major deltas debouch into receiving bodies of water that are large enough for various types of surface waves to be created within the water body. No single factor plays a greater role in coastline development than wave regime. The major effects of waves are to sort and redistribute the sediments debouched by rivers and to mold these sediments into wave-built shoreline features such as beaches, barriers, and spits. In deltas, the continued rapid introduction of sediments by a riverine source interrupts the normal equilibrium between wave regime and depositional topography. In this way, the resulting geometry of many deltaic sand bodies depends not only on the magnitude and distribution of wave forces, but also on the ability of the river to supply sediments. Deltaic sand bodies display a spectrum of configurations ranging from those which have been produced solely by the debouchment of the river without interference from wave action to those which reflect complete dominance by the waves in redistributing the riverine sediments and straightening the coastline. Sand bodies in the former case are normally oriented at high angles to the shoreline trend; whereas those produced primarily by wave action are oriented parallel to the shoreline trend. Whether the sand bodies reflect fluvial or wave dominance depends largely on the ability of the waves to rework and redistribute them.

High nearshore wave power (greater than 20×10^7 ergs/sec) is commonly associated with steep concave offshore slopes and normally sand bodies resulting from this wave energy produce clean, well-sorted permeable sands. Under extreme wave power the mineralogy of the river sands is often changed drastically, always trending towards a higher quartz content in the sand bodies. In some instances, the sands will be completely reworked and only quartzose sediment will be concentrated along the shoreline. Low nearshore wave power (less than 1×10^7 ergs/sec) is commonly associated with low convex offshore profiles. Sand bodies formed under such conditions are the products of riverine processes and the sand bodies generally trend at high angles to the shoreline. Sands are normally poorly sorted and clay-bound.

Application of a comprehensive quantitative wave climate program to a large number of major deltas indicates that deltaic configurations and coastal landform combinations depend to a considerable degree on the wave power adjacent to the shoreline and on the river discharge relative to wave forces. Nearshore wave power is not correlative with deep-water wave power, but is in large part controlled by frictional attenuation which is the function of the subaqueous slope. The subaqueous slope in turn depends partially on the width of the continental shelf, and primarily on the rate at which the river can supply sediments to the nearshore zone. River dominated configurations result only when the river is able to build flat offshore profiles which reduce nearshore wave power; where the subaqueous slope is steep, waves reach the shore comparatively unmodified and wave-built forms such as beach-dune ridges prevail.

Morphological characteristics of seven major river deltas are summarized in Table 2.1 and their mean annual discharge and wave power climate indexes are given in Table 2.2. The ratio of the average discharge per unit of channel width to the average nearshore power per unit of crest width is referred to as the discharge effectiveness index. The absolute value of this ratio has no physical meaning; however, the relative values and ordering provide a significant means of comparing deltas in terms of the relative degree of riverine or wave dominance. The higher the index, the greater the ability of the river to deliver sediment to the shoreline unimpeded by wave action; the lower the ratio, the greater the dominance of extreme wave energy over the delta. In the latter case wave processes contribute significantly to reworking of the river-derived sediments. The attenuation ratio shown in Table 2.2 indicates the extent to which wave power is lost through friction between deep water and the shoreline. When comparing the delta morphologies in Table 2.1 with the discharge effectiveness index and attenuation ratios in Table 2.2, the morphologies of the river deltas are to a considerable degree functions of river discharge and the strength of the wave forces near the delta shoreline. In addition, the wave power near the shoreline is not proportional to the wave power in deep water; it depends primarily on the subaqueous profile, which drastically attenuates the waves. Thus, many of the earlier classifications of river deltas based upon deep-water wave power appear now to be invalid. Before a river can effectively oppose the sea to develop a river-dominated deltaic configuration, it must first build a flat, shallow offshore profile to attenuate incoming waves. The actual subaqueous profile that accumulates in front of a particular delta depends partially on the regional slope of the continental shelf but also on the rate at which the river can supply sediments to the

Table 2.1. Characteristic Delta Morphologies

Coastline and river mouth configuration	Delta shoreline landforms	Delta plain landforms
Mississippi		
Highly indented coastline, multiple extended digitate distributaries—"bird-foot"	Indented marsh coastline, sand beaches scarce and poorly developed	Marsh, open and closed bays
Danube		
Slightly indented with protruding river mouths	Marsh coastline with sand beaches adjacent to river mouths	Marsh, lakes, and abandoned beach ridges
Ebro		
Smooth shoreline with single protruding river channel	Low sand beaches and extensive spits with some eolian dunes	Salt marsh with a few beach ridges
Niger		
Smooth, arcuate shoreline, multiple river mouths slightly protruding	Sand beaches nearly continuous along shoreline	Marsh, mangrove swamp, and beach ridges
Nile		
Gently arcuate, smooth shoreline with two slightly protruding distributary mouths	Broad, high sand beaches and barrier formation with eolian dunes, beach ridges at distributary mouths	Floodplain with abandoned channels and a few beach ridges, hypersaline flats and barrier lagoons near present shoreline
Sao Francisco		
Straight, sandy shoreline with single slightly constricted river	High, broad sand beaches with large eolian dunes	Stranded beach ridges and dunes
Senegal		
Straight coastline with extensive barrier deflecting river mouth	High, broad sand beaches with large eolian dunes	Large linear beach ridges and swales, eolian dunes

Table 2.2. Mean Annual Discharge and Wave-power Climate Indexes. Wave-power values are given per centimeter of wave crest. The discharge effectiveness index is normalized to maximum discharge.[a,b]

River	Wave power (erg/sec) Deepwater	Nearshore	Discharge rate x 10^3 (m^3/sec)	Discharge Effectiveness index	Attenuation ratio
Mississippi	1.06×10^8	1.34×10^4	17.69	1.00	7913.3
Danube	2.30×10^7	1.40×10^4	6.29	2.14×10^{-1}	2585.0
Ebro	7.28×10^7	5.09×10^4	0.55	4.87×10^{-2}	1299.5
Niger	6.76×10^7	6.59×10^5	10.90	8.03×10^{-4}	102.8
Nile	1.36×10^8	3.21×10^6	1.47	5.86×10^{-4}	42.5
Sao Francisco	3.71×10^8	9.97×10^6	3.12	2.37×10^{-4}	37.2
Senegal	1.56×10^8	3.77×10^7	0.77	4.75×10^{-5}	4.2

[a] Wave-power values are given per centimeter of wave crest.
[b] The discharge effectiveness index is normalized to maximum discharge.

nearshore zone. Figure 2.5 shows the average subaqueous river profile of the seven rivers whose characteristics are shown in Tables 2.1 and 2.2. The Mississippi River has an extremely low offshore profile (Figure 2.5). The average offshore slope is approximately 7% (0.12°). This low offshore profile drastically attenuates the incoming deep-water waves and results in extremely low wave energy along the shoreline. As a consequence, the delta shoreline of the Mississippi is characterized by highly indented marshy coastline where sand beaches are scarce and poorly developed. The Senegal River Delta has a much higher offshore slope (Figure 2.5), the offshore slope averaging 17% (0.40°). In this instance deep-water waves are not modified significantly and the shoreline is characterized by extremely high wave action. The delta shoreline is characterized by a wave-straightened shoreline, and high, broad sand beaches are characteristic landforms throughout this delta plain.

Wave forces are the primary mechanism, therefore, whereby the sea reworks and molds deltaic sediments into nearshore marine sand bodies. This particular process plays a highly significant role in controlling the geometry and distribution of sand bodies, their orientation and mineral composition.

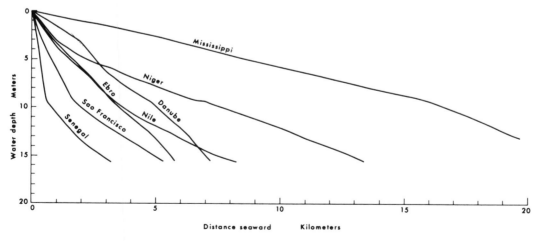

Figure 2.5. Average subaqueous slopes of seven river deltas.

6. Tidal Processes

Many rivers debouch into high tide range environments where tidal processes play a major role in producing deltaic sand bodies. A few notable examples include the deltas of the Ord (Australia), the Shatt-al-Arab (Iraq), the Amazon (Brazil), the Ganges-Brahmaputra (Bangladesh), the Klang (Malaysia), and the Yangtze-Kiang (China). In those deltas dominated by tidal action, at least three important characteristics can be identified: (1) water mass mixing by tidal activity destroys vertical density stratification so that the effects of buoyancy at the river mouths are negligible; (2) for a portion of the year, in some instances the larger part of the year, tides account for the highest percentage of the sediment transporting energy, and flow both in and seaward of the river mouths is subjected to reversals over a tidal cycle that cause bidirectional sediment transport; (3) the zone of marine-riverine interactions is greatly extended both vertically and horizontally.

Tidal amplitudes in macrotidal rivers tend to be large relative to water depths at the river mouth, and the tidal wave in the lower channel is significantly deformed. This shallow-water tidal deformation in tide-dominated channels causes the velocity of flood currents increasingly to exceed that of the ebb currents with distance upstream. In and seaward of the mouths of macrotidal rivers, bidirectional tides commonly rework the sediments supplied by the river into linear subaqueous sand ridges. These ridges have been described by Off (1963) and Wright et al. (1975). These large ridges parallel tidal currents and separate zones of flood and ebb dominated sediment transport. The tidal ridges are illustrated in Figure 2.3e. The pronounced tidal asymmetry in which average flood velocities significantly exceed average ebb velocity results in tide-induced bed shear stresses that produce a net upstream effect over the bed as a whole; hence there is appreciable upstream transport of bedload. The largest and most prominent bed forms found in the lower distributaries are thus flood oriented. Only in the deeper water portions of the channel do bed forms tend to show an ebb dominated direction. This upstream transport of bedload by flood tide currents causes extensive sand accumulations within the channels and in many instances the vast majority of the cross-bedding directions will show an upstream orientation, although bidirectional cross-bedding can be found in some cases. In addition, as a channel becomes abandoned by the river it will normally be sand-clogged by the continued activity of tidal currents. Thus sand-filled channels are of major significance in tide-dominated deltas.

Macrotidal channels are also characterized by bell-shaped river mouths that are commonly choked by the sandy shoals described above. Flood-dominated or upstream-bedload transport augments point-bar growth and necessitates lateral channel migration in order that a channel of sufficient depth be maintained. Quite often, intense meandering patterns are found just upstream of maximum tidal influence, and well-defined sand bodies tend to develop within the deltaic plain. This pattern contrasts sharply to channel patterns in river deltas where tidal range is low. Here the velocity in the distributary channel is normally high enough to keep most of the sediment supplied to it in suspension or moving along the bottom as bedform migration. Therefore, little sediment accumulates in the channel, and most channels display U-shaped cross sections with few shoals. After abandonment of the channel, in most instances, these channels are filled with fine-grained deposits and contrast sharply to the sand-filled channels of macrotidal river deltas.

Macrotidal range deltas also display considerable overbank crevassing and crevasse splays. During flood tide, current velocities and maximum turbulence are attained when the channel is at mid-tide level and large amounts of sediment are put in suspension. The presence of a strong flood tide combined with a high tidal range causes a backwater effect upon seaward-flowing riverine water, and very commonly the river will top its banks, causing extensive crevasse splaying. As the tide level peaks, the water begins to flow out of the banks, velocity drops to zero, and the sediment is deposited as overbank splays. As the water level then drops, velocity again increases; it attains maximum velocity at mid-tide level and the ebb phase. At low tide the velocity again drops to zero, and large amounts of sediment are now flushed out and deposited at the river mouth bar, forming large accumulations of sand at the river mouth.

The movement of tidal waters in and out of the channel and over the lower delta plain is also of basic importance to depositional patterns in the interdistributary and distributary margin regions. Intricate networks of tidal channels and creeks are common features of many deltas having high tidal range environments. In tropical humid regions where abundant vegetation is present, the tidal plain is often characterized by extremely heavy growth of mangroves and other salt-tolerant plants. The vegetation tends to stabilize and maintain the intricate tidal channel network. Many of these tidal channels tend to be characterized by sand-filled bidirectional bedload transport, and many of the sands display sharp bases and resultant high cross-bedding. In more arid macrotidal regions, interdistributary flats are composed of silts and clays deposited from suspension in overbank flows, and in many instances these crevasse splays interfinger with evaporite layers that form in the more interior portions of the interdistributary regions.

Tidal processes, which to a large extent place strong control on cross-bedding of channel sands, produce intensive overbank crevassing, cause channels to be sand-filled in high-tide regions, are responsible for accumulation of large, linear, sandy tidal ridges seaward of the river mouths, and result in the formation of scour-base channel sands by the migrating tidal channels commonly found in interdistributary regions. In most instances, deltas that experience extreme tidal range display primarily tidal flat environmental characteristics, and these sediments differ considerably from deltaic facies that form in lower tidal ranges.

7. Wind System

In delta plains where large, low coastal plains are persistent, the wind regime plays a significant role in the formation of subaerial deposits. The wind is responsible for eolian transport of sediment over the subaerial currents and the setup or setdown of the water surface along the coast. In many regions, and more commonly in arid regions, these processes are of critical importance for the development of various morphologic landforms in the coastal plain. In deltas where persistent onshore winds and high wave energy are present, the entire subaerial delta deposit will consist primarily of wind-blown sand which is worked back over the deltaic plain as a transgressive eolian sheet. Wind stress applied to nearshore waters creates local wind waves, which in many cases are responsible for reworking sediments and concentrating small strandline sand bodies, especially in the vicinity of the river mouths. This process is common in those deltas fronted by large, low-gradient continental shelves, which commonly attenuate deepwater waves, but the unlimited fetch results in wind waves reaching considerable energy levels. The small sand beaches normally associated with the river mouths of the Mississippi River are the result of reworking of deltaic sediments by local wind-generated waves. Wind-generated waves also help to keep fine-grained sediment in suspension once it has left the river mouth, and currents spread this material laterally, leading to a progradation of the coastal plain downdrift of the delta proper.

The most significant influence of wind systems is in producing nearshore coastal currents both at the surface and along the bottom. In delta regions where tidal forces are extremely low, wind stress is commonly responsible for driving currents capable of moving coarse-grained sediments. Strong onshore winds will cause a piling up of water along the coastline (setup). Surface currents follow the wind shoreward, and bottom-hugging currents trend offshore or at high angles to the coastline. Persistent offshore winds, on the other hand, create the opposite effect; water level is lowered along the coastline (setdown) and surface currents move offshore while bottom currents trend onshore or parallel to the coastline. Persistent alongshore winds normally drive currents in the direction of the wind. Such currents are significant agents in transporting sediment away from the river mouths and quite often result in the formation or accumulation of broad expanses of mud in the downdrift direction from the delta proper. Such accumulations play major roles in the progradational sequence of many deltaic facies.

8. Littoral Currents

Littoral currents are driven by several forces: deep oceanic currents impinging against the continental margin, tidal propagation, winds, waves, and density gradients. The Guiana Current, an oceanic current that flows northward along the front of the Amazon Delta, transports the heavy sediment-laden waters northwest, and much of the sediment debouched by the Amazon has been deposited considerable distances (800 to 1,000 km) away from its active river mouths. Much of the sedimentary material accumulated as mud-flat and beach-ridge coastlines of northern Brazil, French Guiana, Surinam, and British Guiana is composed primarily of material derived from the Amazon.

Tide-driven coastal currents are often strongly asymmetrical, resulting in a dominant longshore transport of sediment away from the river mouths. Such currents are often found in regions experiencing higher tidal ranges or narrow seaways. The tidal currents in the Malacca Straits set primarily to the northwest, dispersing riverborne sediment northward, away from all of the river mouths debouching into the Straits. These riverine sediments are transported in an alongshore direction and commonly accumulate downdrift of the major delta as broad muddy or sandy tidal flat deposits. If appreciable wave action is present, the tidal flats are commonly sandy and well sorted and form broad sand sheets marginal to the delta. If wave action is low the sediments accumulate as poorly sorted sandy silts and clays which are normally characterized by a high faunal content, and only occasionally will small, linear, sandy beach ridges (cheniers) be present within these deposits.

Wind-wave induced nearshore currents are of primary importance in low tidal regions and partially enclosed basins. In many instances these currents develop into nearshore circulation cells that are often associated with the formation of the rhythmic topography commonly seen in many offshore bars. This type of circulation results in the formation of large-scale low-angle cross-bedding in nearshore sands. Across many continental shelves wind-driven currents can often attain velocities that can move and concentrate coarse material in offshore settings. Many of the sand sheets accumulating marginal to active deltas have formed as a result of sediment transport by wind-driven currents. During the passage of storms and hurricanes bottom currents, driven primarily by high wind velocities, can transport sediment far offshore, thus concentrating coarse lag deposits at considerable water depths. In addition, the stress induced by strong bottom currents across many muddy shelves can result in producing various types of deformational features within the sedimentary column and in some cases are responsible for large-scale subaqueous slumping processes.

The major role played by all types of currents is in orienting sand bodies subparallel or parallel to depositional strikes. Quite often at considerable distances downcurrent or offshore from the active delta, directional properties produced by currents often parallel the long axis of the sand body, and quite often the sand body displays low-angle, large-scale internal cross-bedding.

9. Shelf Slope

High rates of sediment accumulation and rapid progradation associated with river deltas result in lower slope angles than those found in most other coastal environments. Slopes off the modern deltas vary from a low of 0.003° off the Parana River to as high as 0.48° off the mouth of the Senegal River. It is this subaqueous portion of the continental shelf that primarily influences frictional attenuation of incoming deep-water waves and therefore can be regarded as a control that determines the nearshore wave power. Thus the continental shelves fronting many modern deltas are purely depositional in nature and are formed in a rather short period of time; most of these shelves show direct evidence of progradation during modern times. This situation contrasts to that of shelves fronting other coastal environments where, in many cases, the shelf deposits are largely residual or lag accumulations formed during Pleistocene sea-level fluctuations or are of tectonic origin.

The slope of the continental shelf plays an important role in determining the pattern of delta switching which happens over geologically longer periods of time. Figure 2.6 depicts three types of delta migration patterns. The first type consists of lobe switching, a condition in which the delta progrades as a series of distributary channels and after a period of time abandons the entire system and forms a new lobe in adjacent regions (Figure 2.6:I). Very commonly this new lobe occupies an indentation in the coastline between previously existing lobes that through time and subsidence have allowed encroachment of the sea. In this manner successive deltaic lobes overlie one another, forming stacked, multiple regressive sand sequences. This type of switching pattern is found most commonly where offshore slope is extremely low, wave power is low, and tidal range is generally below 2 m. In many instances the distributary mouth bar deposits associated with each delta lobe merge one with another to form major sheet-type sands. The second type of delta switching consists of major shifts of the channel far upstream in the deltaic plain and a corresponding

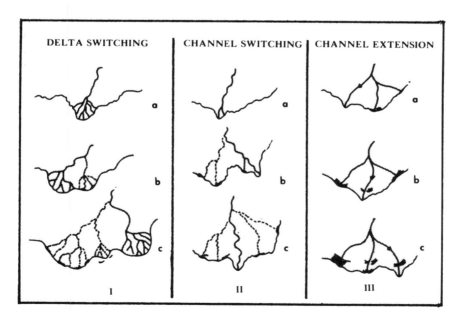

Figure 2.6. Types of delta switching patterns.

new course exists for the river and its delta (Figure 2.6:II). The Hwang Ho and Ganges-Brahmaputra Deltas are modern-day examples. This type of pattern is characterized by intermediate shelf slopes, high persistent wave energy, and a high tidal range. The third type of delta switching is referred to as alternate channel extension (Figure 2.6:III); the Danube and Red River Deltas display this form. Two or more major distributaries break off at a nearly common point at the head of the delta and continue unbranched to the river mouth. Commonly one of the distributaries will carry the majority of the sediment-water discharge at any given time. As a result, this active channel will rapidly prograde seaward, but the other distributaries show little progradation and commonly are being wave reworked along their margins. Eventually the prograding distributary will lose its gradient advantage by overextending itself, and the discharge will seek one of the other, shorter distributaries. With the increased sediment down the new channel, it will rapidly prograde, leaving behind a series of stranded beach ridges. This process will be repeated multiple times, forming a deltaic plain characterized by multiple sequences of beach ridges.

The topography of the continental shelf also plays an important role in controlling the rates of delta formation and progradation. Submarine canyons near the mouths of modern deltas such as the Ganges-Brahmaputra, Indus, Congo, and Sao Francisco direct sediments offshore, where they often form large deep-water submarine fans. The modern subaerial delta of the Ganges-Brahmaputra is one of the largest in the world, yet the submarine fan at the base of the canyon fronting this delta is several times larger than the subaerial delta and has accumulated in extremely deep water. Regardless of the origin, these canyons play an important role in tapping the sediment discharged by modern river systems and funneling it seaward across the continental shelf and slope into deep-water areas.

10. Tectonics of the Receiving Basin

The three-dimensional aspect of deltaic sequences is in large part controlled by the stability of the depositional site. Although this parameter is significant, it is virtually impossible to quantify in any fashion. In some deltas, however, enough data have been gathered to show the effects of rapid subsidence and in others the relatively shallow depth to bedrock indicates stability. However, in the vast majority of the deltas studied there is a definite lack of information to ascertain the nature of basin stability. Rapid subsidence in most regions is caused by both tectonic downwarping and shallow dewatering of clays by overlying denser deltaic sediments. In many instances rearrangement of the facies by penecontemporaneous deformation (diapirism, growth faulting, and subaqueous mass movement) is common in many deltas that debouch large quantities of fine-grained suspended sediment onto the continental shelf. In all cases rapid subsidence produces a significant thickening of deltaic sands, the thickness commonly increasing several times over that of normal deposits. In those receiving basins that display low subsidence rates and relatively high stability, thin deltaic sequences are more common. Sand bodies are normally widespread and tend to show high lateral continuity. Undoubtedly this parameter is much more significant than indicated in this short discussion; however, the general lack of data precludes any quantitative treatment.

11. Receiving Basin Geometry

The geometry of the receiving basin appears to exert strong control on delta configuration and delta switching patterns. Existing tectonic maps and available data indicate that five types of basin geometry are apparent in the river systems studied. These five types of generalized configurations are graphically illustrated in Figure 2.7. The first type consists of a narrow trough open at both ends which normally provides a passage connecting two larger bodies of water (Figure 2.7:I). Modern-day examples include the Straits of Malacca, the Sea of Japan, and the Mozambique Channel. Structural downwarping has occurred within the basin, and sediments delivered by the river are fed from the sides of the basin into the central axis of the seaway; thus river flow or sediment introduction is perpendicular to the basin axis. The narrow restriction causes tidal currents to be extreme, and many of the deltas display skewed alongshore configurations and are tide and current dominated.

The second type of basin geometry consists of a closed-end, narrow structural trough, with the river supplying sediment at the closed end of the trough (Figure 2.7: II). This sediment is dispersed and deposited along the axis of the trough. Modern-day examples include the Gulf of California, the Bay of Bengal, the La Plata Estuary, and the Persian Gulf. In most instances, the tidal wave is significantly modified by the basin shape, and currents tend to move in and out of the narrow, elongate basin rather than displaying strong alongshore components. Quite often the lower ends of the deltaic distributaries are totally tide dominated. Large, linear, offshore tidal ridges previously described are commonly associated with the river mouths of deltas prograding into this type of basin. The third type (Figure 2.7: III) consists of a downwarped area that lies inland of the shoreline yet the site of active deltaic deposition is seaward of this point and progrades onto a rather stable platform. The Niger and the Yangtze-Kiang River systems are typical of this type of delta debouching into this basin geometry. In such instances large inland freshwater deltas and large swampy areas are common within the upper parts of the deltaic plain and the alluvial valley. As the rivers cross these broad, rapidly subsiding inland basins they lose much of their sediment load prior to reaching the coast. In most instances the thickest sequence of deltaic deposits is in the inlandmost regions rather than at their seaward margin. An area of active subsidence that lies seaward of the present shoreline with a depositional basin that is essentially open ended constitutes the geometry of the fourth type of basin configuration

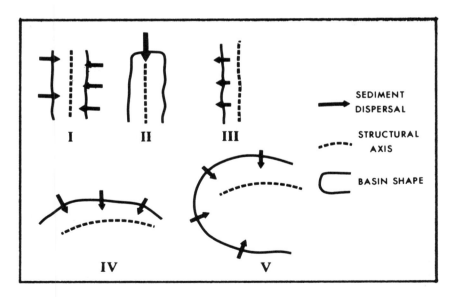

Figure 2.7. Major configurations of receiving basins.

(Figure 2.7: IV). These types of basins are normally found fronting the major open ocean basins. Modern-day examples include the Grijalva, the Sao Francisco, and Senegal River Systems. In such instances the delta itself will never fill the basin.

The fifth type of basin geometry (Figure 2.7: V) is characteristic of semi-enclosed or enclosed basins, such as the Gulf of Mexico, Caspian Sea, and Black Sea. In most instances, a zone of rapid subsidence lies close to one side of the basin. Rivers emptying into this type of basin commonly display features characteristic of rapid continuing subsidence. In many instances the basin edge will be dominated by one or more larger river systems with numerous minor river systems debouching onto the basin margin between the larger controlling river systems.

12. Summary

Various interacting dynamic processes acting on sediment are responsible for producing the wide variety of deltaic sand body geometries and distribution that exist in modern deltas. The following list summarizes the major control that each process discussed in the previous section exerts on deltaic facies.

I. Climate
1. Controls sediment-water yield;
2. Controls in situ delta deposits; tropical—large, thick accumulations of peat; temperate—thin, high, laterally continuous peat layers; arid—complex interfingering supra tidal and evaporite deposits.

II. Water Discharge Regime
1. Erratic discharge regimes produce braided channels displaying wide lateral continuity.
2. Nonerratic discharge regimes produce more stable meandering channels (shoestring sands).
3. Erratic discharge regimes commonly result in numerous complex interfingering fining-upward sequences displaying highly variable porosity-permeabilty relationships.

III. Sediment Yield
1. High fine-grained suspended load rivers build extensive unstable platforms characterized by complex compactional and deformational features.

IV. River-Mouth Processes
1. Distributary-mouth bar sand body geometry and distribution controlled by three major forces—inertial (narrow, linear sand bodies); buoyant (thin, widespread, coalesing sand bodies) and frictional (bifurcated channels with middle ground sand bar and capped by subaqueous natural levee deposits).
2. All distributary mouth bar deposits display coarsening-upward sequences.

V. Wave Power
1. High, persistent wave power produces straight delta shorelines; sand bodies display marine characteristics and are oriented parallel to depositional strike or

form clean, well-sorted sheet sands; sand bodies display high quartzose content independent of parent material.
2. Low wave power results in irregular indented delta shorelines and sand bodies oriented at high angles to depositional strike; sand bodies often clay-bound and poorly sorted.

VI. Tidal Processes
1. High tidal range deltas have sand-filled channels; numerous sandy overbank crevasse splays; large, complex meander-belt sand bodies in upper delta plain.
2. Macrotidal ranges in narrow seaways, straits, etc., result in formation of major linear offshore tidal ridges and bars that often attain lengths of tens of kilometers and thicknesses that approach 20 to 30 m.

VII. Wind Processes
1. Directional variability in coastal wind systems often imparts multidirectional cross-bedding to coastal sand dunes.
2. In microtidal region wind stress on nearshore water masses often controls currents that shape and orient offshore sand bodies.

VIII. Nearshore Currents
1. Sediment transporting currents in delta regions are driven by several forces: permanent deep oceanic currents impinging on the shelf, tidal propagation, wind and wave driven and complex density currents.
2. Currents are responsible for orienting offshore sand bodies parallel or subparallel to depositional strike; sand bodies are often located considerable distances offshore or down-current from active delta lobe; sand bodies commonly display strong marine bedding characteristics.

IX. Shelf Slope
1. Subaqueous slope controls offshore wave dissipation, which strongly influences sorting and hence porosity of nearshore sand bodies.
2. Low offshore slopes commonly associated with multichannel distributary patterns which can form complex sand body relationships.
3. Shelf topography, especially presence of submarine canyons, plays major role in topping delta sediments and formation of deep marine sand bodies.

X. Tectonics of Receiving Basin
1. Rapidly subsiding basins result in overthickening and localization of deltaic sand bodies, whereas relatively stable basins display widespread and laterally continuous delta sand facies.
2. Localized differential weighting and dewatering of sediments in receiving basin results in formation of large-scale penecontemporaneous structures and sediments often display evidence of subaqueous mass sediment movements, displaced sediment, and complex slumping.

XI. Receiving Basin Geometry
1. Large-scale geometry of receiving basin exerts significant control on delta configuration and delta switching patterns.
2. The large-scale geometry of most receiving basins results in formation of similar delta relationships for large distances in an alongshore direction.

3 | MISSISSIPPI RIVER DELTA

1. Introduction

The Mississippi River, the largest river system in North America, drains an area of 3,344,560 km². The average discharge of the river at the delta apex is approximately 15,360 cumecs (cu m/sec) with a maximum and minimum of 57,900 and 2,830 cumecs, respectively. Sediment discharge has been estimated by various investigators, but is generally about 2.4×10^{11} kg annually. The sediments brought down by the river to the delta consist primarily of clay, silt, and fine sand, the clay load accounting for a high percentage (70%) of the total sedimentary load.

Figure 3.1 shows the drainage basin and the alluvial valley of the Mississippi River Delta [note the relationship of the size of the delta (28,568 km²) in relationship to the extremely large drainage area of the river]. The drainage basin of the Mississippi River Delta consists of a wide variety of igneous, metamorphic, and sedimentary rocks and thus, the sediment load is composed of a highly varied mineralogical suite. Since at least Cretaceous time, the drainage basin of the Mississippi River system has been delivering sediments to the Gulf of Mexico. Through time, the sites of maximum deposition, or depocenters, have shifted within the Gulf Coastal Plain. The bulk of the sediments that comprise the Gulf Coast geosyncline have been derived in part from ancestral Mississippi River drainage. Thus, the river system has been operative over long periods of geologic time, constantly feeding sediments to the receiving basin, building up a thick sequence of deltaic sediments which, since Cretaceous time, have prograded the coastal plain shoreline seaward.

In recent times, the seaward progradation of the Mississippi River deltas has constructed a deltaic plain which has a total area of 28,568 km² of which 23,900 km² is subaerial in nature. During the past 7,000 years the site of maximum sedimentation or delta lobes has shifted and occupied numerous positions. Figure 3.2 illustrates the generalized location of these delta lobes during the past 7,000 years. One of the earlier deltas, the Sale Cypremort, prograded and constructed a deltaic lobe along the western flanks of the present Mississippi River delta plain. This delta lobe was extremely widespread and generally thin (10 to 15 m thick). After approximately 1,200 years of buildout the river system lost its gradient advantage and the site of maximum deposition switched to another delta lobe, the Cocodrie system (Figure 3.2). A similar sequence of events ensued and with time, this site of active deposition was abandoned and a new delta lobe (the Teche) began a period of active

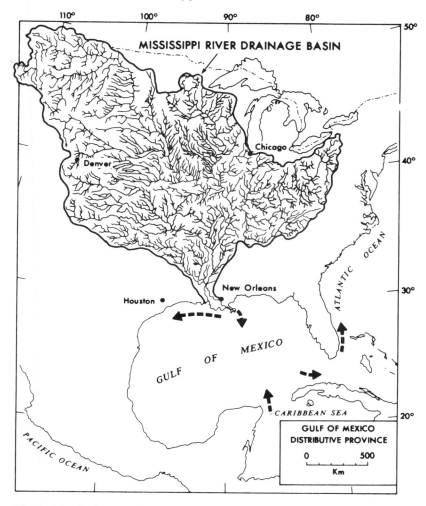

Figure 3.1. Drainage basin of the Mississippi River.

buildout. This process continued until the present location of the modern delta was established some 600 to 800 years ago. The phenomenon of progradation and subsequent abandonment of individual delta lobes has resulted in the formation of the offlapping and laterally displaced deltas. The net result is a complex of deltaic sequences extending approximately 300 km along the Louisiana coast and inland nearly 100 km.

The phase of buildout or progradation of the deltas has been termed the constructional phase by Scruton (1960). Once the river abandons its rapidly constructed delta, the unconsolidated mass of deltaic sediment is immediately subjected to those processes associated with marine reworking and subsidence. In deltas experiencing rapid subsidence marine waters will immediately encroach over the regressive sequence of deltaic sediments. Marine processes such as waves, tides, and currents will have only a minimum amount of time in which to rework the sediments of the regressive phase. The subsequent transgressive sequence of deposits over the abandoned delta will consist primarily of shallow marine and bay sediments; the regressive phase will be preserved in its entirety. In other types of settings in which subsidence is not extremely rapid, wave attack along the seaward margin of the delta will begin to erode, transport, and redistribute the sediments of the uppermost

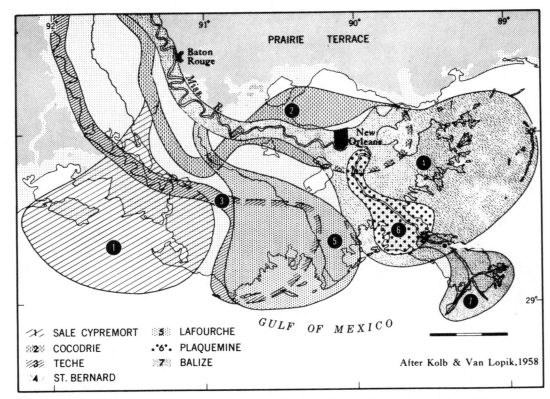

Figure 3.2. Deltaic lobes of Mississippi River deltas. (Modified from Kolb and Van Lopik, 1958.)

portions of the regressive phase. The net result is the destruction of the upper portion of the delta mass and the winnowing and concentration of sands along the seaward margins of the abandoned delta. This phase of the delta sequence has been termed the destructional phase by Scruton (1960). An excellent example of the destructional phase is illustrated by the St. Bernard Delta (Figure 3.2). Some 600 years ago the St. Bernard Delta ceased to actively prograde seaward. With the diminished influx of sediments, seaward progradation ceased and subsidence and wave reworking became the dominant process. The subaerial delta, deprived of overbank sedimentation, began to open up into small bays and lakes, and with continued subsidence the lakes and bays began to merge with one another. An encroachment of marine waters slowly began on top of the old delta plain. Along the seaward margins of the delta, waves and currents reworked the tip ends of the prograded distributaries. The fine-grained material was re-suspended and carried away from the delta proper by littoral currents. Sands and silts were concentrated by wave processes and a sand barrier island began to form along the seaward margins of the delta. Offshore from the barrier island thin sand sheets began to accumulate. Each sand sheet occurs in an arcuate belt approximately 120 km long and has widths on the order of 8 to 14 km. Thickness of the sand sheet is highly variable, but occasionally sands will accumulate to thicknesses of 7 to 10 m. In the initial stages of reworking the shoreline is rapidly pushed landward by the constant reworking action of the waves and currents. With time, however, a considerable amount of sediment is concentrated by these processes and landward retreat slows down appreciably. During this period of decreased erosional rates, subsidence is continuing and the major portion of the delta plain begins to subside below sea level. In the St. Bernard Delta, a large interior sound or lagoon has formed (Chandeleur Sound) (Figure 3.2). Thus along the seaward edges of the regressive phase, transgressive sands will cap progradational sequences. Inland from this point, however, shallow marine, bay and lagoonal deposits will overlie the prograded facies. Continued subsidence of the deltaic mass will eventually result in complete inundation of the St. Bernard Delta regressive phases by marine waters, and thus the site one day will again be occupied by another progradation of a Mississippi River delta lobe.

This orderly repetition of depositional events and shifting sites of sedimentation result in numerous interfingering and overlapping regressive deltaic sequences separated by shallow water marine deposits or transgressive sands. The initiation, growth, and abandonment of the various delta lobes result in cyclic alternations of detrital and nondetrital deposits. Each major regressive lobe is comprised of a detrital lens or complex of lenses, generally bounded on all sides by essentially nondetrital sediments indigenous to the basin of deposition. The detrital lens or the regressive phase is characterized by a high percentage of relatively coarse clastics, sands and silts, abrupt facies changes, and rapid accumulation and burial rates. The bounding sediments, which tend to be generally richer in organic constituents and chemical precipitants, show a slower depositional rate and tend to be tabular accumulations with considerable lateral continuity. The cyclic concept provides a framework for organizing the complex environmental relationships and facies distributions resulting from delta building.

2. Environments of Deposition

The environments of deposition associated with the regressive or progradation phase of deltas are extremely complex and highly variable. In most river deltas there is a constant sediment supply of a given grain-size distribution and composition during the period of progradation. The various processes active in each subenvironment of the delta distributaries result in selective concentration and distribution of the sediments. These processes described previously result in the formation of recognizable environments of deposition within a given delta.

The subenvironments of the regressive phase can generally be divided into two major categories. Those associated with the subaqueous delta consist of sediments that accumulate beneath sea level and those subenvironments associated with the subaerial delta consist of the deposits that accumulate at or above sea level. Subenvironments associated with the subaqueous delta consist of: (a) shelf, (b) prodelta, (c) delta front, (d) distributary mouth bar, and (e) subaqueous natural levees. The subenvironments associated with the subaerial environment consist of: (a) channel deposits, (b) natural levee or overbank splays, (c) interdistributary bays, (d) marshes, and (e) crevasse splays or bay fills.

2.1 Shelf Environment

The continental shelves fronting the Mississippi Delta are generally of low gradient and consist of deposition of fine-grained clays transported from the delta proper out into the deeper waters of the continental shelf. These fine-grained deposits are delivered to the shelf via several mechanisms. During periods of extremely high wave action, turbulence associated with wave processes will maintain suspension of the sediments within the water column. Current stresses carry these sediments away from the active delta and allow normal vertical deposition of the fine-grained clays onto the shelf proper. By this process clays generally tend to be widespread away from the active delta and sedimentation blankets much of the continental shelf. Sedimentation rates are slow (a few millimeters per century).

A second process is the delivery of fine-grained clays by a process associated with internal wave activity. Near the mouths of the active distributaries of the river, density interfaces between the overlying fresh waters and the underlying saline marine waters are extremely sharp. Commonly along these interfaces, internal waves will be generated. The internal waves tend to propagate from the river in a seaward direction. Often associated with the formation and propagation of these internal waves are zones of flocculated clays which by the process of seaward migration of internal waves carry considerable amounts of fine-grained sediments away from the river mouth and continental slope of the Mississippi River Delta. Details of this process have been described by Wright and Coleman (1971, 1974).

A third major process resulting in deposition on continental shelves results from various types of mass movements, slumping, and sediment flowage associated with normal delta progradation. This latter process probably contributes the greatest amount of sedimentation on the continental shelves seaward of most river deltas and can result in a high influx of sediments originally deposited high on the delta flanks and subsequently, via mass movement processes, allows the sediments to be pushed out onto the continental shelf. This process will be described in detail in a later section.

The resulting shelf deposits generally are grouped into three major types of sedimentary accumulations. The first type of shelf facies is a product of extremely slow accumulation of fine muds generally removed from areas of active progradation. Distribution and ultimate deposition of the fine-grained clays is controlled by wind-driven currents and semi-permanent alongshore currents. Cores taken in this type of shelf facies are typically composed of apparently homogeneous clay, shelly clay or silty clay. Shells and shell fragments are scattered throughout the deposits. Bedding is normally defined by differences in color and in kind and frequency of authigenic inclusions (shell fragments, manganese oxide nodules, organic fibers, etc.). Contact between the stratifications is normally sharp and well-defined. Parallel and lenticular laminations are relatively rare due to the intense activity of burrowing organisms. Individual laminae of silt and fine sand, when present, are normally distinctive and sharp basal boundaries. Occasionally the silt layers are graded, normally from coarse at the base to fine at the top. One of the most characteristic aspects of these deposits is the intense burrowing. Quite commonly the burrows are filled with foraminiferal tests, small shells, and shell fragments. Secondary deposits of pyrite and siderite will commonly line the burrow cavities. The lithology as well as the structural assemblage generally indicates a slow deposition under marine conditions and in an environment where the most active process of sediment modification is burrowing by bottom-living organisms. Occasionally currents and strong wave action agitate and rework the bottom, resulting in the thin silt lenses.

The second type of shelf facies results when the offshore region adjacent to the deltas receives little or no clastic accumulations. Organic debris, precipitants, and reworked material become the dominant lithology. Notable features include the concentrations of shell, foraminifera tests and other calcareous material as well as large well-defined burrows. This facies will be characterized by widespread accumulation of shell concentration or hashes ranging in thickness from 5 cm to a meter or more, and show a dominance of organic over detrital deposition. These shell hashes commonly display wide lateral continuity and are overlain and underlain by clay rich deposits displaying abundant burrows and shell concentrations. Very commonly extending below the shell hashes are zones of large well-defined burrows. The burrow fills are characteristically derived from the shell hash above.

A third type of shell facies consists of sediments deposited on the shelves via numerous subaqueous mass movement processes. These deposits characteristically display little or no internal bedding. Massive clays and silty clays showing little or no internal structure often alternate with finely parallel bedded silts and clays deposited via vertical sedimentation on the continental shelf. The massive, subaqueous, mass-moved, transported sediments vary in thickness ranging from a few meters to over 15 to 30 m. Often sand bodies slumped from higher positions on the delta front are incorporated in these massive shelf deposits.

2.2 Prodelta Environment

The basal portion of the active prograding delta is commonly referred to as prodelta deposits. This facies in relationship with other facies of the subaqueous delta forms an easily recognizable stratigraphic sequence. The facies relationship of these subaqueous delta environments is shown diagrammatically in Figure 3.3. The prodelta facies is characteristically a blanket of clays deposited from suspension having high lateral continuity and low lithologic variation. In many instances, the transition from shelf deposits into the prodelta facies can be distinguished only when the associated subaqueous delta complex is known either in a vertical or a plan view sequence. Because deposition is entirely from suspension, parallel laminae are by far the most common primary structure. In many instances the prodelta deposits will normally appear as structureless massive clays; however, X-ray radiography tends to reveal that parallel laminae are present and are defined either by inclusions of diagenetic origin or by differential sedimentation rates and consequently slightly differing textural associations. Because of the higher rates of deposition associated with the prodelta deposits, normally these deposits will escape intense burrowing by marine organisms. In its most seaward portions the prodelta deposits display extremely thin laminae normally consisting of color banded clays. Burrowing organisms will be present occasionally and alternate with laminae that display little or no burrowing. Faunal species are normally high, especially formaniferal tests, indicating an open marine environment; both number and species are high. The shallower water portions of the prodelta deposits tend to show laminae that are thicker and grain size is normally coarser. Parallel and lenticular silt laminae become more common and tend to

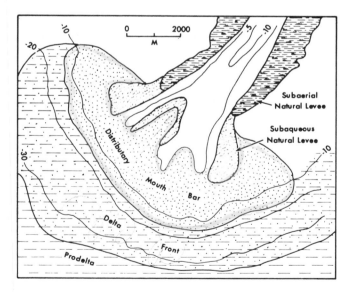

Figure 3.3. Depositional environments at a river mouth.

34   *Deltas: Processes and Models*

replace the color laminae of the lowermost zones of the prodelta. Cores of the prodelta environment displaying typical characteristics are illustrated in Figure 3.4.

The thicknesses of the prodelta deposits vary considerably from delta to delta. In the Mississippi River Delta these deposits can assume considerable thicknesses ranging from 20 to 100 m. In other deltas, where fine-grained deposits are not abundant, the prodelta facies generally display much coarser textures; silty clays and silts will generally tend to be thinner in nature (generally less than 20 m).

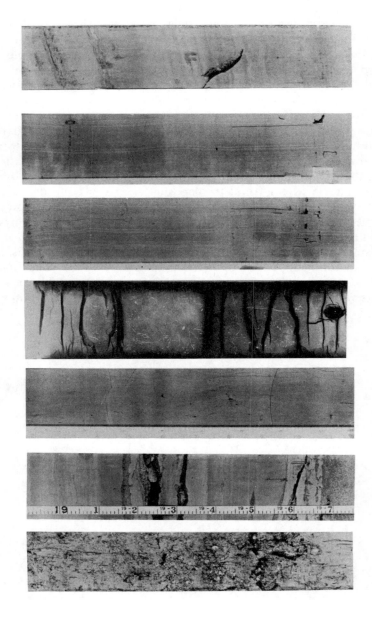

Figure 3.4. Cores of Mississippi River prodelta deposits.

2.3 Delta Front Environment

Progressing upward or landward within the deltaic sequence, the deltaic front (also referred to as the delta platform or distal bar) overlies the prodelta facies. This environment is the seaward sloping margin of the advancing delta sequence. Increase in sedimentation rates and coarseness of the sediments distinguishes these deposits from the prodelta clays. Lithology generally can be characterized as laminated silts and clays with thin sand layers. The depositional slope within this environment is high in comparison with most other deltaic environments. However, rarely do slopes on the Mississippi Delta exceed one-half to three-quarters degree. The delta front generally shows lower lateral continuity than the prodelta environment (Figure 3.3).

In the more seaward portions of the delta front deposits small burrows and shell remains are scattered throughout the deposits and often result in partial destruction of the parallel and lenticular laminations. In some instances the lower portion of the delta front is characterized by intense zones of burrowing organisms. Dilutions of salinity by river water along with the nutrient-laden currents derived from the river make this a highly favorable environment for burrowing organisms, especially polychaete worms, and often cores taken in the more seaward portions of the delta front are characterized by high bioturbation. Because of the shallower nature of these deposits and the near proximity to the river mouths, currents associated with high floods on the river will normally feel bottom and produce a wide variety of sedimentary structures associated with current and wave processes. Sedimentary structures such as small-scale cross-laminae, current ripples, scour and fill, and erosional truncations attest to these physical processes. In the uppermost portions of the delta front environment, sequences commonly display alternating beds of silt and silty sands, all displaying well-developed textural laminae. Typical cores from the delta front environment are illustrated in Figure 3.5.

2.4 Distributary Mouth Bar Environment

The distributary mouth bar is an area of shoaling associated with the seaward terminus of a distributary mouth (Figure 3.3). Shoaling is a direct consequence of a decrease in velocity and a reduction in carrying power of a stream as it leaves the confines of its channel. Accumulation rate is extremely high, probably higher than in any other environment associated with the delta. The sediments are constantly subjected to reworking, not only by stream currents but also by waves generated in the open marine water beyond the channel mouth. Detailed documentation of the processes associated with the development of the distributary mouth bar have been described in numerous articles (Scruton, 1960; Gould, 1970; Fisk, 1961; Coleman et al., 1964; and Wright and Coleman, 1974) and these discussions will not be repeated here.

A general understanding of the processes and mode of formation of the distributary mouth bar is critical to understand the evolution and vertical relationships in a deltaic sequence. Figure 3.6, modified from an original diagram by Scruton, 1960, illustrates the process of sediment dispersal at a river mouth. Less dense fresh river water flows via the distributary mouth over denser saline gulf waters. The coarser sediments, the sands, settle rapidly from suspension and almost all of the sand is deposited near the vicinity of the distributary mouths. Because of variations in turbulence at the river mouth and different process intensities between low river stage and high river stage, silts and clays will also be deposited in this environment. Thus in many instances, the distributary mouth bar can contain and incorporate silts and clay beds. However, reworking by marine processes (predominantly during low water periods) results in cleaning and sorting of the sediments

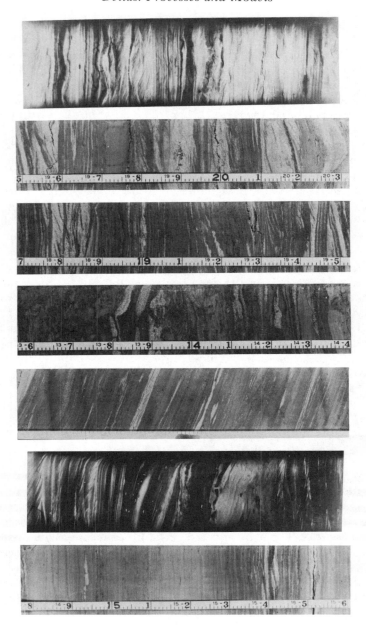

Figure 3.5. Cores from the distal-bar deposits.

by marine processes. As a result, the distributary mouth bars consist commonly of clean well-sorted sands delivered by the river to its mouth. The remaining finer grain suspended load carried by the river is distributed widely by the expanding river effluent which is constantly decreasing in velocity seaward. Both suspended load and bottom sediments become progressively finer with increased distance from the distributary mouths. As the sediment load is constantly diminishing and being spread in an ever increasing area, the rate of clastic deposition gradually decreases seaward from the distributary mouths. The coarser-grained sediments carried both as bed load and in suspension are deposited near the

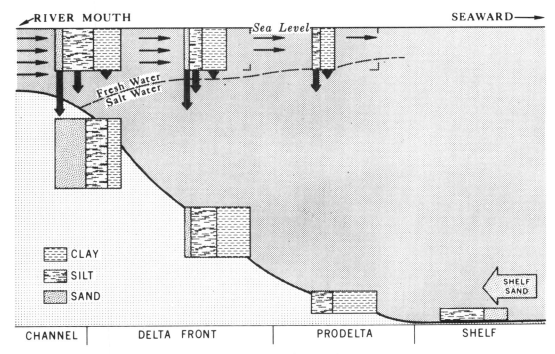

Figure 3.6. Sediment dispersal at a river mouth. (From Scruton, 1960.)

river mouth and the distributary mouth bar (Figure 3.6). Further seaward of this region, the finer-grained components predominate as one moves progressively seaward through the delta front environment and finally into the prodelta deposit *furthest* from the river mouth.

Figure 3.6 illustrates the development of the facies in a single instant in time. However because of the continued influx of sediment to the distributary mouth over a hydrologic year, these facies prograde seaward with time. Figure 3.7, modified from Gould (1970), shows the progradation of the distributary mouth bar at Southwest Pass during the period 1764 to 1959. During this two-hundred year period the distributary prograded 10 km seaward. The lower section of Figure 3.7 illustrates this process in cross-sectional perspective. This dispersal process combined with the continued progradation results in the formation of the typical coarsening-upward sequence associated with all delta sequences and forms the bulk of the regressive phase. Under such conditions, the lithologic facies form a near horizontal lithofacies as one environment progrades over another environment of deposition. Time lines, however, represent depositional surfaces and tend to cross the lithologic boundaries (Figure 3.8). At times in the past the depositional surface showed the same profile as the present day but it was located further landward in a position indicated by the dashed time lines. With progradation, the surface advanced seaward to its present position because of deposition at the river mouth. An understanding of this relationship is fundamental to interpreting the environments of deposition associated with the subaqueous delta.

The deposits of the distributary mouth bar consist primarily of the coarsest clastic material available in the delta system. In the Mississippi River, the distributary mouth bars are composed of fine-grained sands that normally display good sorting, and high porosity and permeabilities. The most common sedimentary structure consists of a variety of types of small scale cross-laminae and current ripple drift. Slight differences in sorting of the distributary mouth bar sequence results in local variations in sand content within the

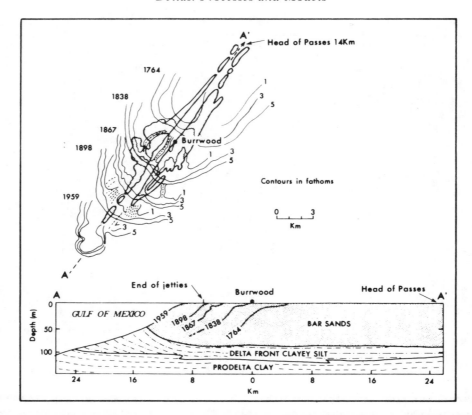

Figure 3.7. Seaward migration of distributary mouth bar at Southwest Pass, Mississippi River, during the period 1764–1959. (Modified from Gould, 1970.)

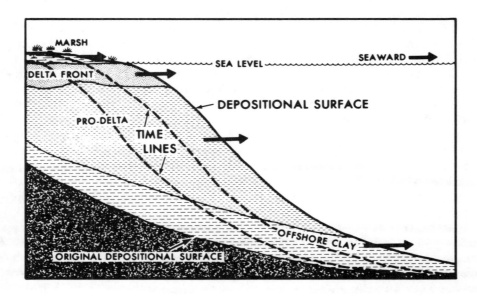

Figure 3.8. Time lines and lithofacies in a prograding distributary. (Modified from Scruton, 1960.)

overall deposit. Near the top of the unit in rivers that drain basins in temperate or tropical climates, large accumulations of river-transported organic debris are found often. Water-saturated logs and other organic debris are transported down the river in times of flood and discharged into the nearshore zone where wave action grinds down the coarser wood particles into concentrations of large quantities of organic debris in the upper portions of the distributary mouth bar. In the Mississippi River Delta, local concentrations of this wood debris are often referred to as coffee grounds. The accumulation of such transported debris would result (after burial and diagenesis) in the preservation of channel or transported coals within the sand body. Such coals, although of high quality, will not normally display extremely wide lateral continuities and thicknesses can change considerably over short distances. In the modern distributary mouth bars of the Mississippi, such organic accumulations can attain thicknesses on the order of 2 to 5 m and show lateral continuities generally less than 1 km.

2.5 SUBAQUEOUS NATURAL LEVEE ENVIRONMENT

Submarine ridges bordering the channel form in response to reduced current velocity as the channel broadens and shoals. These submarine ridges are commonly referred to as subaqueous natural levee deposits. In the Mississippi River Delta the distributary mouths are dominated by friction-dominated effluents which issue into a very shallow offshore bottom. Initially, the rapid rate of effluent expansion characteristic of this type of river mouth produces a broad arcuate radial bar. However, as deposition continues, natural subaqueous levees develop beneath the lateral boundaries of the expanded effluent where velocity gradients are steepest. The formation of these natural levees tends to inhibit any further increases in effluent expansion rate, so that with continuing bar accretion, continuity can no longer be satisfied simply by increasing the effluent width. As the central bar continues to grow upward, channelization takes place on the threads of maximum turbulence which tend to follow the edges of the subaqueous levees. The result is the occurrence of a bifurcating channel which has a triangular middle ground shoal separating the divergent channel arms. The edges of the channel are thus characterized by the accumulation of sediments that form the subaqueous natural levee ridges.

The bulk of the subaqueous levee facies is composed of very fine sand and silt with occasional thin laminae of plant debris or clay. Current-produced structures are the dominant features of this environmental assemblage. The combined action of current and wave processes yields many complex forms of cross laminations. One of the most common sedimentary structures associated with the subaqueous natural levee is climbing ripple drift. Other associated sedimentary structures are illustrated in typical cores from the subaqueous natural levee illustrated in Figure 3.9.

2.6 SUBAERIAL NATURAL LEVEE ENVIRONMENT

The areas of slightly higher elevation bordering and confining the channel are called subaerial natural levees or overbank splays. These ridges form in response to differences in river stage. Deposition is solely the result of overbank flow. Highest elevations and broadest widths are found where river stage fluctuation is greatest. The distributary natural levees attain higher elevation upstream and slope seaward at an average rate of 0.1 m/km. In the lower delta of the Mississippi, the subaerial natural levees range in elevation from a maximum of 2 m along major distributary channels to those that occur virtually at sea level near the mouths of the distributaries.

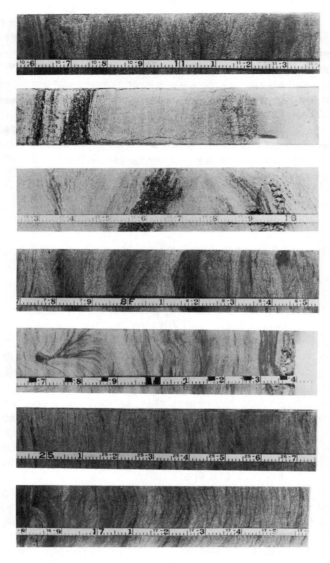

Figure 3.9. Cores of Mississippi River distributary mouth bar deposits.

During annual floods on the river, the river stage will normally top the adjacent natural levees. However, flood inundation is not continuous along all portions of the river and in many instances during a single flood crevasse splays will form the small lobes emanating along low points of the natural levee. These small crevasse splays will be maintained for a year or several years until they build up or aggrade the natural levee to the point of the flood level and then they will cease to be active. This continuous process of differential overtopping of the channel margins results in the buildup and formation of the natural levees or crevasse splays. Figure 3.10 shows a low oblique aerial photograph of an overbank crevasse splay formed along a distributary of the modern Mississippi River Delta. The general discussion of these crevasse splays, their modes of formation, and resulting sedimentary characteristics can be found in Arndorfer (1973). Typical cores taken in the

Figure 3.10. Overbank splay in lower Mississippi River Delta.

natural levees or crevasse splays are illustrated in Figure 3.11. These deposits are characterized by a wide variety of sedimentary structures (Figure 3.11). Climbing current ripples in a variety of small scale cross-laminae are very common in this facies. In many respects they resemble the environments of the subaqueous natural levee and the delta front. One distinguishing characteristic, however, is the presence of intense root-and-worm-burrowed bioturbations on the upper surface of bedding planes. In addition, exposure to subaerial oxidizing conditions commonly results in the formation of a large number of diagenetic products, especially iron carbonates and iron oxides. Often the subaerial natural levee deposits will be capped by a tightly cemented zone of iron oxides and iron carbonates that form in conjunction with the highly bioturbated zones. This facies is normally not widespread and generally is confined to parallel bands following the distributary channels.

2.7 Distributary Channel Environment

A distributary channel is a natural flume which accommodates and directs a portion of the discharge and transported sediment from the parent river system to the receiving basin. In most instances within the lower delta, the distributary channels are rather stable and do not show a tendency to migrate laterally, thus preventing the formation of point bar or meander belt deposits. In the present delta, active channels range in size from less than 2 to 3 m wide with depths on the order of 1 m to those as large as a major river, 1 km wide to 30 m deep. Depth generally decreases rapidly as the river mouth bar is approached and water depths over the bar rarely exceed 3 m.

Regardless of the size, processes within the channel environment are essentially similar. In the upstream part of the channel, current flow is confined by banks to a general downstream direction. Beyond the mouth, however, the channel flares and currents

Figure 3.11. Cores of Mississippi River subaerial natural levee.

frequently move in varying directions. With continued progradation of the distributary channels, a point is reached at which the channel is no longer able to maintain its gradient advantage and the process of channel abandonment begins. Deprived of an active influx of sediment and water, the lower portions of the channel are commonly filled with poorly sorted sands and silts containing an abundance of transported organic debris. As the waters in the channels become more stagnated and lower current velocities are maintained, the finer-grained materials begin to infill the channel proper. With time and continuing subsidence, often the channel is entirely filled with the fine-grained, poorly sorted sediments. Fine-grained organic debris and logs commonly form the upper part of the channel fill. Thus, in the Mississippi River Delta where tides are low, there is no process by which to infill the channels with sand or other coarse material; the channel fill deposits are commonly lenticular bodies of fine-grained material interfingering locally with peats and transported organic debris. Figure 3.12 shows a low aerial oblique photograph of an abandoned channel in the modern Mississippi River Delta. This particular channel is Balize's Bayou, a major distributary of the Mississippi River in the latter part of the 18th century. At the turn of the 19th century the channel began to receive less volume of water and sediment and the process of abandonment began. Today, the entire channel is only a small poorly defined channel representing the final stages of infilling of this channel (Figure 3.12). Borings through this channel indicate that the sedimentary fill consists of a lower portion characterized by poorly sorted sands and silts commonly showing an abundance of organic trash along the bedding planes. In the central portion of the channel fill, highly bioturbated silts, silty clays and clays characterize the fill sequence. Localized

Figure 3.12. Photograph of an abandoned channel in the Mississippi River Delta. The fill is composed of organic rich clays, silts, and peats.

slump features, indicated by high dips in the strata, intraformational clay clasts and the occasional presence of large plant debris, such as limbs and logs, are common. In the uppermost portion of the channel fill, fine-grained organic debris, primarily peats, with abundant woody horizons are common; consequently, the final stage of filling is normally characterized by a high abundance of peat.

2.8 INTERDISTRIBUTARY BAY ENVIRONMENT

The interdistributary bay environment includes areas of open water within the active delta which may be completely surrounded by marsh or distributary levees but which are more often partially open to the sea or connected to it by small tidal channels. These are generally shallow water bodies rarely exceeding four meters in depth containing brackish to marine waters. The interdistributary bays are more commonly elongate with their longest dimension ranging in size from a few hundred meters to approximately 15 km. The more elongate bays flanking the outer sides of the natural levees have been called levee flank depressions by Russell (1936).

Because these bays are generally bypassed by active coarse clastic sedimentation, the deposits are composed primarily of fine-grained sediments that are brought into the bays during periods of high flood or abnormal high tides associated with the passage of storms. The most abundant single sedimentary structure in the bay deposits consists of lenticular laminae, a product of reworking and concentration of the coarse fraction by wind generated waves. The horizontal extent of these small layers varies but is generally less than 50 cm. Parallel laminae are occasionally encountered and consist of alternating zones of fine silt and silty clay. It is believed that these laminae are formed during times of high flooding of the Mississippi River. Current ripple marks and scour-and-fill structures are common in some cores and indicate that currents are occasionally active during times of deposition.

Probably the most common stratification consists of a wide variety of bioturbation by organisms. The faunal remains are commonly abundant and, in many cases, developments of small oyster reefs can be found within the bay deposits. Fauna generally indicate a shallow brackish water marine environment. One of the more characteristic features of the bay deposits is the inclusion of abundant sporomorphs within the deposits. Samples from the active interdistributary regions of the modern delta show an extremely high content of reworked spores and pollen ranging in age from Cretaceous through the modern. Figure 3.13 shows typical cores from the interdistributary bay deposits of the Mississippi River Delta.

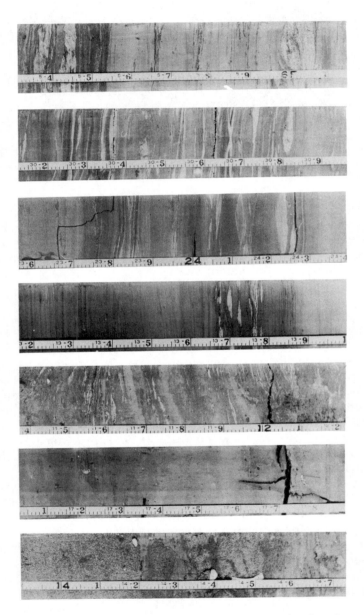

Figure 3.13. Cores of Mississippi River interdistributary bay deposits.

2.9 Crevasses or Bay Fill Environment

One of the major facies associated with many deltas, including the Mississippi, is the large aerial extent of the bay fill or crevasses that break off of the main distributaries and infill the numerous interdistributary bays. These sequences form the major land areas of the subaerial delta and form as crevasse deposits in shallow bays between or adjacent to major distributaries and extend themselves seaward through a system of radial bifurcating channels similar in plan to the veins of a leaf. Figure 3.14 illustrates the bay fill sequences that have formed within the modern delta during the past few hundreds of years. Of the six crevasses shown in Figure 3.14, four have been dated historically and much of their development can be traced by historic maps. Each bay fill forms initially as a break in the major distributary natural levee during flood stage, gradually increases in flow through successive floods, reaches a peak of maximum deposition, wanes, and becomes inactive. As a result of subsidence, the crevasse system is inundated by marine waters reverting to a bay environment, thus completing its sedimentary cycle. The mass of sediment resulting from the process of crevassing is relatively thin, from 3 to 15 m in thickness; this sequence formed in a period of approximately 100 to 150 years. Although individual bay fills are relatively thin, continuing subsidence and a repeating of similar processes result in stacking one bay fill on top of another, thus building up quite a thick sequence of subaerial delta deposits.

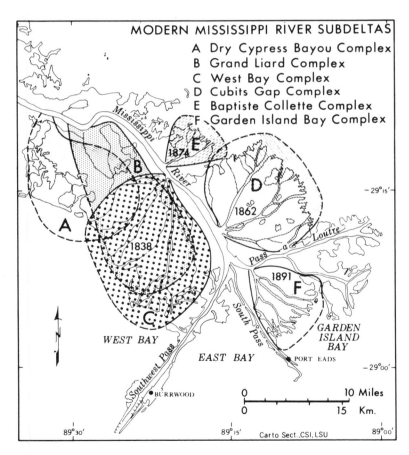

Figure 3.14. Crevasses or subdeltas of the modern Birdfoot Delta of the Mississippi River.

A plan view of an idealized crevasse system is shown in Figure 3.15. The crevasse initiates as a break in the major distributary levee in the vicinity of point A. During the early formative years, buildout was characterized by coarse deposition in the immediate vicinity of the break. With time new channels formed, bifurcated and reunited, forming an intricate pattern of distributaries. Later in its history, some distributaries became favored while others were abandoned and became inactive. With the development of systematic channel patterns the delta front advanced into the bay resulting in the deposition of a sheet of relatively coarse detritus thickening locally in the vicinity of the channels. Seaward of the active river mouths is an area of fine-grained accumulation or prodelta clays. Other parts of the crevasse system which have become abandoned or deprived of sediment nourishment undergo rapid subsidence and many areas tend to open up and revert back to shallow marine bay environments. In cross section, the prodelta clays constitute the basal portion of the detrital lens (Figures 3.16a, b). These clays mark the first introduction of detritus into the bay and are distinguished from the underlying shallow marine clays by alternate repetition of sedimentary structures related to seasonal floods. These distinct annually deposited laminae increase in thickness from the bottom to the top within the prodelta clays, faithfully recording the approach of the locus of active deposition towards a point in the bay. The uppermost portion of the detrital lens is composed of silts and sands deposited immediately in front of the advancing river mouth. The detrital lens, therefore, consists of two major facies and shows distinctive geometric form related to the radial bifurcating system of channels originating from a point source. The size distribution within this lens grades from fine to coarse, both upwards in the section and horizontally from the distal ends of the system towards the point source.

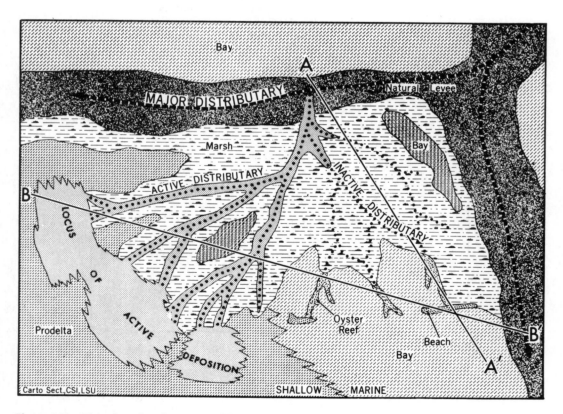

Figure 3.15. Plan view of environments of deposition in a crevasse. (After Coleman and Gagliano, 1964.)

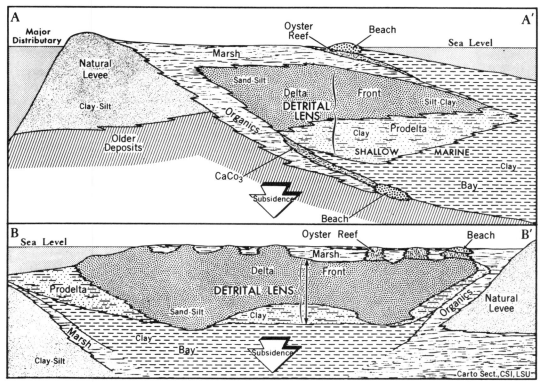

Figure 3.16. Cross sections of a crevasse. Location of cross sections shown on Figure 3.15. (After Coleman and Gagliano, 1964.)

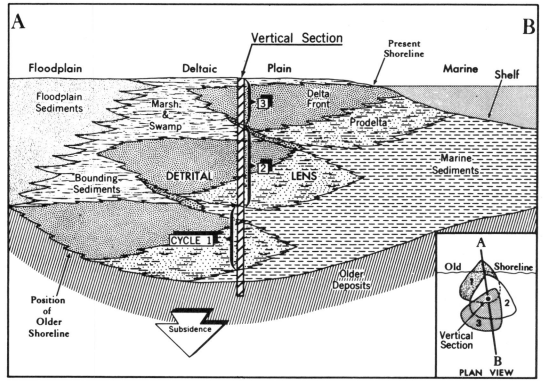

Figure 3.17. Idealized delta cycles resulting from delta progradation (After Coleman and Gagliano, 1964.)

Once active sedimentation ceases in the bay fill, subsidence and coastal retreat become the dominant process. For a time, marsh growth can keep pace with subsidence, but eventually large bays tend to develop and the shoreline retreats rapidly. Locally small beaches accumulate in the vicinity of the natural levees where a coarser grain of material is available for reworking. These beaches are usually thin and discontinuous. Oyster reefs may find a foothold along the old channel margins of submerged levee ridges. Thus the regressive sequence is capped by bounding transgressive sediments including shallow marine bay clays and marsh deposits. These deposits contain a higher percentage of elements characteristic of the basin. The initiation, growth, and abandonment of the various bay fills result in cyclic alternations of detrital and nondetrital deposits. These cycles occur on many scales ranging in thickness from those only a few meters thick to those that exceed 10 to 15 m in thickness. The aerial extent of an individual cycle is also highly variable, but 400 km^2 is a good approximation of many of the crevasses associated with the modern delta.

Figure 3.17 illustrates some of the possible facies arrangements resulting from overlapping bay fill deposits; also, it illustrates a simple case in which three bay fills have prograded seaward, each overlapping the previous one. The relationships of one cycle to another are complex; however, they are orderly in that the sequence of development of each lobe is similar. In vertical section (Figure 3.17), three complete cycles are recorded and different facies relationships are represented in each cycle. For example, in cycle 1 only marine prodelta clays represent the detrital lens and shallow marine bay deposits of the bounding sediments are present. In cycle 2, both prodelta and delta front deposits are encountered in the detrital lens and the bounding sediments are composed of clays, peats, and oyster reefs. The detrital lens of cycle 3 advanced well beyond the point of the vertical section and only the coarsest portion is recorded; the bounding surface is entirely nonmarine. Therefore, the history of seaward progradation is well documented in this section.

The previous examples illustrate hypothetical developments of the bay fill sequence. However, many historic examples exist in the modern delta. Figure 3.18 illustrates the development of the West Bay subdelta or bay fill. The diagrams represent tracings from historic maps; only a few of the large number of maps are reproduced in the diagram. The break in the levee occurred in 1839 and at that time maps in the vicinity of West Bay showed water depths on the order of 7 to 10 m. After the initial break in the levee, coarse sediment was deposited subaqueously in the immediate vicinity of the break and no new land developed. However, with continued deposition, and a general shoaling of the bay in the immediate vicinity of the break, a channel pattern rapidly developed and this organized distributary pattern rapidly filled the bay. This stage of development is shown in Figure 3.18b at which time there were numerous active channels developed and the entire bay had been nearly filled with bay fill sediments or the regressive sequence of delta progradation. By 1922 many of the channels had been abandoned; they had prograded far enough seaward to lose their gradient advantage and only a few active channels were continuing to deliver sediments to the Gulf of Mexico. A luxuriant marsh growth had covered the entire bay, and peats were forming on top of the regressive sequence. The most active processes at this time were the production of organic material by the luxuriant plant growth attempting to keep pace with the rapidly subsiding basin. Eventually plant growth could no longer maintain its pace and subsidence became the dominant process. Slowly the marsh broke up into numerous small lakes and bays. Wind-generated waves on the shallow bays began to rework the upper portions of much of the regressive wedge of sediments. By 1961 much of the original buildout had subsided beneath sea level and numerous open bays were apparent on the maps (Figure 3.18d). By 1976, nearly the entire region had been inundated by marine waters and the West Bay complex had reverted back to a shallow marine bay environment. Given time, eventually another crevasse would form on the banks of the Mississippi and

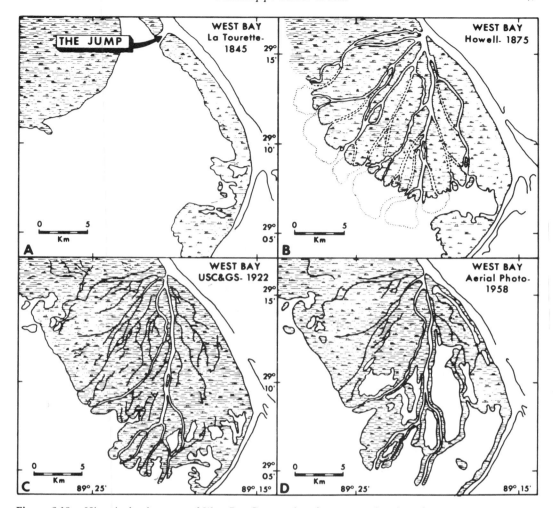

Figure 3.18. Historic development of West Bay Crevasse based on maps of various dates.

another period of progradation would ensue, again filling the West Bay with detrital sediments. It is this process of repeated filling, alternating with periods of subsidence and reverting to open bay conditions, that forms the bulk of the cyclic deposits that cap most subaqueous delta sequences. This particular process is responsible for forming the bulk of the upper delta deposits.

An extensive sedimentological study was conducted of the West Bay subdelta region in 1961 by the Coastal Studies Institute of Louisiana State University. Over 200 continuous cored borings, normally 18 m in depth, were drilled in the West Bay region. In addition to the cored borings, some 300 drill holes were completed and thus an excellent three-dimensional aspect of the bay fill was reconstructed. Figure 3.19 shows the location of the cored borings completed in this region; whereas Figure 3.20 shows a fence diagram of the reconstructed facies relationships of this particular region. Up to three complete cycles of progradation and subsequent abandonment can be distinguished (Figure 3.20). The most widespread units present in the region consist of the shallow marine bay clays and the peats which cap and separate the detrital cycles. These deposits form the bounding deposits which encase the detrital regressive wedge of the bay fill. At the point of crevassing, the

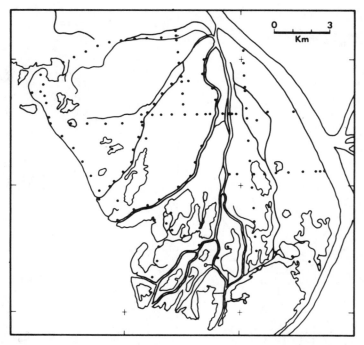

Figure 3.19. Core hole location map in West Bay Crevasse. These core holes provide the data for the fence diagram illustrated in Figure 3.20.

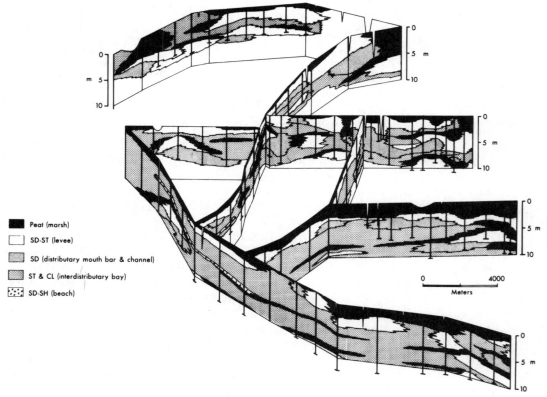

Peat (marsh)
SD-ST (levee)
SD (distributary mouth bar & channel)
ST & CL (interdistributary bay)
SD-SH (beach)

Figure 3.20. Fence diagram of delta facies in the West Bay Crevasse or bay fill.

sands of the detrital lens are extremely thick and generally tend to erode or cut down into an earlier cycle. In the more central portion of the fill, the cycles are generally complete in nature. That is, each cycle begins with a marine clay representing the original bay bottom, grades upward into the prodelta and delta front deposits which consist of silts and clays. This in turn is overlain by the coarsest of deposits of the regressive sequence which represents the distributary mouth bar. Finally, the capping deposits of the marsh cover the regressive sequence. In some portions of the fence diagram numerous complete cycles can be recorded. In some instances, the cycles are separated by thin transgressive sand deposits which represent local reworking of the earlier cycle and concentrations of sands and shell material into a thin beach deposit. In such bay fills, however, beach deposits are normally not common because of the lack of intense wave reworking and the rapidity at which the process from regression to abandonment takes place.

Marine clays predominate in the seaward-most areas of the fence diagram (Figure 3.20). Most of the regressive sequences did not prograde to this point in the West Bay area and thus incomplete cycles are recorded. The most common vertical sequence along the distal ends of the bay fill is represented by shallow marine bay deposits which grade upward into silty clays representative of the seaward-most portion of the prograded bay fill. This silty clay deposit is normally capped by a thin peat deposit such as seen in the fence diagram. The peats in this vicinity are generally thin, but tend to show wide lateral continuity and thus the entire cycle consists only of silty clays and peats that alternate in an orderly fashion.

Figure 3.21 shows a schematic diagram illustrating a lithologic log and a hypothetical electrical log of a completed bay fill cycle. The lowermost sequence in the diagram represents the organic clays and silty clays which represent the initial fine-grained fill of the bay. The sandy deposits represent those facies associated with the actively prograding distributaries and consist primarily of delta front and distributary mouth bar deposits. One characteristic feature is that much of the upper portion of the sandy deposit is characterized by highly root-burrowed sequences. Overlying the distributary mouth bar are the natural levee deposits and the calcareous clays of the small interdistributary bays and overbank deposits. The final lithologic unit is the peat deposits that tend to cap the regressive sequence and thus a complete cycle is recorded.

These bay fill sequences are often conspicuous on the outcrop and form the only coarse clastic detritus in an otherwise monotonous sequence of shales on the face of the outcrop. Figure 3.22 shows an example of an ancient bay fill sequence in the Carboniferous rocks of Kentucky. The unit shown in Figure 3.22a represents the lower portion of the bay fill and

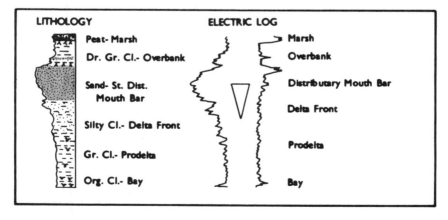

Figure 3.21. Schematic diagram of a bay fill showing electric log response.

consists of highly burrowed shales and silty shales. Scattered fossils are found throughout this interval. The sequence in Figure 3.22b represents the initial part of a bay fill sequence that corresponds to the prodelta clays of the modern bay fill described previously. These deposits display lenticular and parallel laminae with scattered shell debris and zones of burrowing scattered throughout. The facies in Figure 3.22c consists of alternating sands and silts separated by thin shale laminations. This portion represents the upward increase in grain size associated with the advancing delta and is commonly referred to as the delta front. The prominent unit in Figure 3.22d represents the sands of the distributary mouth bar. Tabular units of sand comprise this sequence. Small scale cross laminations can be seen within individual units. The top of this sequence displays highly burrowed and rooted structures representing a buildup to sea level and colonization by marsh plants. The final unit, Figure 3.22e, represents some of the overbank splays and interdistributary bays that cap the regressive phase. The coal (Figure 3.22f), although not shown prominently in this photo, completes the sequence. Above this unit are other cycles and in this particular outcrop there are several cycles one superimposed on top of another.

2.10 Marsh Environment

Marshes are low tracts of periodically-inundated land supporting non-woody plants such as grasses, reeds and rushes. The marsh surface normally approximates the mean high tide level and within the Mississippi River Delta this level is only a fraction of a meter higher than mean sea level. Within the active deltas, marshes generally range from fresh to brackish. The marsh environment's most notable feature is the abundance of plant life and the production and preservation of organic materials. The abundance of organic material coupled with its proximity to the water table affords ideal conditions for the accumulation and preservation of plant material. Most marsh deposits reveal simply the presence of abundant organic accumulations which accumulated under relatively stagnant chemically reducing environments. Occasionally overtopping of the newer banks by major floods will introduce fine clastic clays into an otherwise rich organic region. The end result is production of laterally continuous peat deposits which vary considerably in quality from

Figure 3.22. Outcrop photograph of bay fill sequence in Carboniferous rocks of Kentucky.

point to point. Figure 3.23 illustrates typical cores taken in the marsh environment from the active delta of the Mississippi River.

3. Lateral and Vertical Relationships of Deltaic Facies

The individual environments of deposition described in the previous section show distinguishing characteristics based upon suites of sedimentary structures, grain size, and

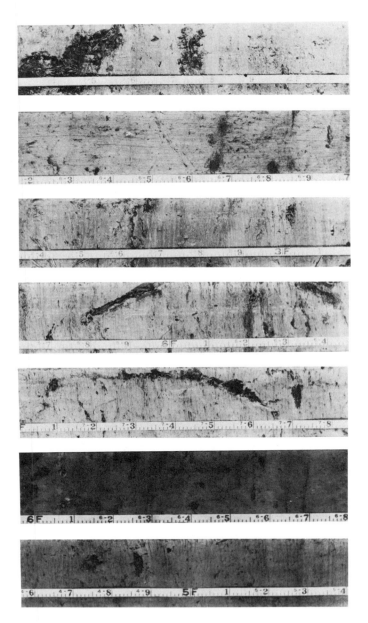

Figure 3.23. Cores of Mississippi River marsh deposits.

faunal content. Thickness and lateral continuity vary considerably from one point in the delta to another depending upon local conditions present at the time of deposition. In subsurface exploration or in outcrop mapping it is often difficult to distinguish an individual environment because of a lack of large enough region of exposure or cores in which to map the details both laterally and vertically of the sequences and their relationship to one another. Figure 3.24 is a block diagram that represents a portion of the Mississippi River Delta in which the distributary channel has prograded beyond the section A-B. During the developmental phase of this portion of the delta, the prodelta deposits built out over existing shelf facies. Interfingering with the delta front deposits are the distributary mouth bar deposits and the associated subaerial delta deposits, the interdistributary bays, the bay fills and the marsh deposits. Note in this particular diagram that the most laterally continuous deposits consist of the more marine type of facies, the shelf and prodelta deposits. The second most laterally continuous sequence is the marsh deposits which tend to cap the advancing delta lobe. The more restricted deposits consist of the distributary mouth bar, the channel fill deposits, the crevasses or bay fills and the interdistributary bays.

Figure 3.25 represents hypothetical electric logs taken on various portions of the block diagram illustrated in Figure 3.24. Locations or positions of these drill holes are indicated on Figure 3.24. The borings A, B, C, D represent the variability that can be obtained as one moves laterally away from the major distributary mouth bar region on which boring D is located. From boring D to boring A, the major sand body, the distributary mouth bar, has diminished in thickness and virtually the entire section consists of marine and prodelta clays with thin delta front deposits being capped by interdistributary bay deposits. The distance between borings A and D could vary considerably depending upon the size of the distributary. In the modern Mississippi River this sand body can extend laterally away from the main channel for distances of 2 to 4 km. Borings E and F, shown in Figure 3.25,

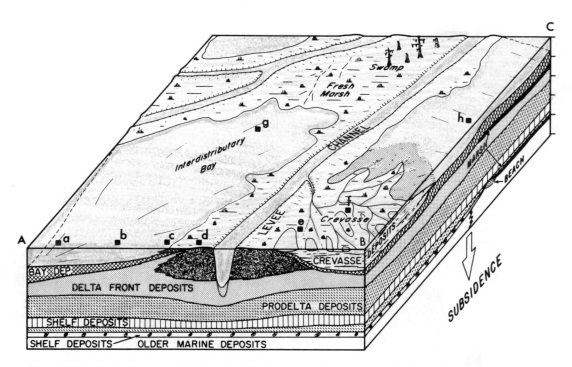

Figure 3.24. Block diagram of deltaic facies. Typical electric log sequences (A-H) are shown in Figure 3.25.

represent sequences that would be obtained in areas adjacent to the main channel and in regions where bay fills make up the largest portion of facies in the entire 75 m section. Boring G is located between two major distributaries and the entire sequence is clayey with the exception of a thin clastic wedge representing the distal portions of the delta front and the bay fill that is shown near the top of the section. At the position of boring H virtually no sands are present within the column with the exception of a small transgressive beach deposit near the base which represents reworking of an earlier deltaic deposit and concentrations of sands by a marine transgression. Very commonly, these sands will display sharp bases and be underlain by peat or coal deposits. With the exception of this sand, the entire section consists predominantly of marine clays and prodelta deposits with a final capping of interdistributary bay deposits which, in this case, can accumulate to considerable thicknesses because of the lack of deposition except by overbank flooding.

The previous diagrammatic examples are based upon a general knowledge of the facies relationships within the modern Mississippi Delta. In several regions of the modern Mississippi, there have been drilled considerable borings for soil foundation studies, and for scientific purposes. Two cross sections (Figure 3.26) are illustrated across the abandoned deltas of the Mississippi River delta lobes. One, AA', is a cross section of the abandoned LaFourche Delta and the second, BB', is a cross section of the abandoned St. Bernard Delta. Core controls on both sections were generally dense with borings normally spaced less than 500 m apart. In both sections, the regressive section of the constructional phase of the delta can be carried laterally for continuous distances. The prodelta clays extend across the entire cross section. The overlying unit, the delta front, also maintains high lateral continuity; however, thickness of the unit varies spatially depending on the proximity of the active distributaries at the time of deposition. The distributary mouth bar, the least laterally continuous unit, is the coarsest deposit present in the section. The subaerial delta is

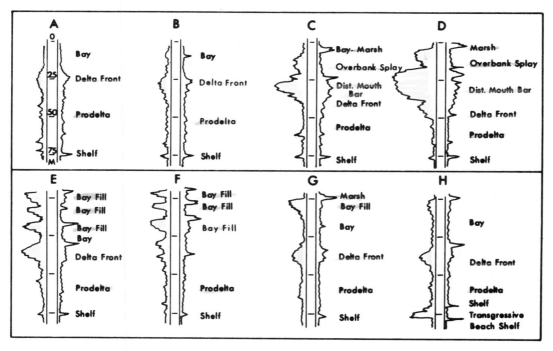

Figure 3.25. Schematic electric log response in deltaic facies. Locations of bore holes indicated on Figure 3.24.

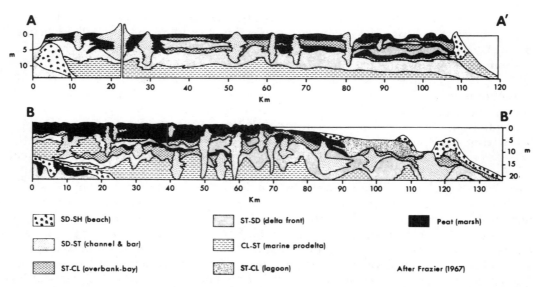

Figure 3.26. Cross sections of Mississippi River delta lobes. Section A-A' runs from N-S across Teche Delta and section B-B' runs N-S across St. Bernard Delta. (Modified from sections presented by Frazier, 1967.)

represented by highly interfingering interdistributary bays, bay fills, overbank splays, and marshes (represented by peat accumulations). The peats show intense interfingering as they thicken and thin with the interdistributary bay deposits. In section BB', which traverses the abandoned St. Bernard Delta, the destruction or transgressive phase of delta development can be seen, and lagoonal or shallow marine deposits overlie the more distal ends of the regressive deltaic sequence. The sand deposits which represent transgressive beach sands overlie the lagoonal deposits and in many cases are scoured through the lagoonal deposits into the underlying delta regressive sands. Because of the rapid subsidence associated with the Mississippi River Delta, the transgressive phase does not entirely erode the regressive phase. The transgression, instead, simply reworks the most distal ends of the delta, concentrates the sand and forms a sheet over the entire deltaic sequence.

4. Subaqueous Deformational Processes: Mississippi River Delta Plain

The hydrodynamic processes operative at the river mouth control the outflow patterns and fundamentally determine the pattern of sediment dissemination and accumulation. Once a sediment has been dispersed and deposited on the subaqueous portion of the delta platform other syndepositional processes involving mass movement begin to modify the depositional patterns and induce a variety of changes in the delta platform. The Mississippi River carries a substantial sediment load annually of which a high percentage consists of fine-grained clays and silts transported as suspended load. The coarser material is deposited at or near the distributary mouths because of the rapid effluent deceleration. The finer-grained sediments are kept in suspension and spread laterally far beyond the immediate mouth of the channel. Figure 3.27 shows a reproduction of an ERTS imagery (Band 5) taken of the modern delta or Birdfoot Delta on November 18, 1973, during a period of low river stage. The suspended sediment plumes (Figure 3.27, light gray color) extend well beyond each river mouth and, even at low river stage, a large mass of turbid water surrounds the active delta. During the flood of 1973, the turbid plume extended distances of 25 to 30 km off the mouth of South Pass, depositing the fine-grained sediment derived from the river beyond the edge of the Continental Shelf in water depths approaching 500 to 600 m. The modern distributaries of the active Mississippi River Delta have extended themselves seaward as long linear channels rather than following the multi-channel bifurcating patterns characteristic of the older delta lobes. This pattern has resulted from the progradation of the modern bird-foot delta across a platform of weak unstable clays and has allowed the distributaries to scour down into marine clays and hence have inhibited active bifurcation as was characteristic of the shallower water earlier delta lobes.

The wide lateral dissemination of fine-grained sediments builds a platform fronting the delta that consists of clays that are rapidly deposited, have an extremely high water content, and, because of the abundant fine-grained organic content, are rapidly degraded by bacterial action producing large accumulations of sedimentary gas (primarily methane and CO_2). Understanding the process of building a fine-grained unstable delta platform is essential to determine the deformational processes that result in subaqueous mass movement.

The modern Birdfoot Delta is the youngest of the delta lobes of the Mississippi River and radio-carbon dates indicate that this delta lobe was formed within the past 600 years. The area of the subaerial bird-foot delta is 1,900 km² compared to an average aerial extent of 6,200 km² of the older Mississippi Delta lobes (each of which has an active life of 800 to 1,000 years). The confinement of the modern delta to a small area has been compensated for by expansion of its vertical thickness. The average thickness of the older delta lobes is 20 m; whereas the Birdfoot Delta is 100 to 120 m thick. Figure 3.28 indicates the typical depositional sequence beneath the modern delta as gathered from hundreds of borings. A complete lithologic faunal and clay mineral analysis of these deposits is described by Morgan et al. (1963). The lowermost units, the strand plain sands, at the base of the sequence are Pleistocene in age and the reefal zone, consisting primarily of cemented algal

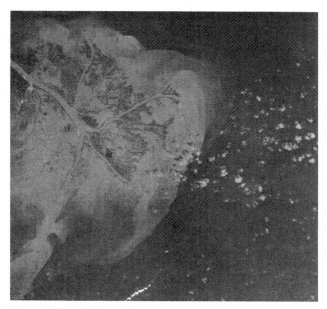

Figure 3.27. ERTS satellite imagery (Band 5) of Mississippi River Delta on November 17, 1973.

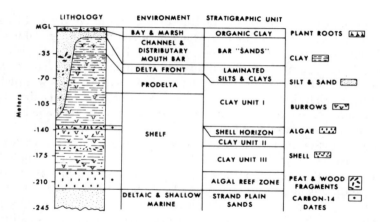

Figure 3.28. Vertical sequence of deposits in modern Mississippi River Delta.

remains, dates at approximately 26,500 years BP (Before Present). Clay units II and III are fine-grained clays of the pre-modern bird-foot delta and are entirely marine in nature. The shell horizon, shown at 130 m, is an excellent stratigraphic marker that can be found throughout the entire delta region. Numerous radio-carbon analyses date this later at 13,400 to 16,600 years BP. Clay unit I is composed of the prodelta clays associated with the modern Mississippi River Delta. The prodelta clay deposits grade upward into the delta front and distributary mouth bar deposits which are overlain by bay fill and marsh deposits of the subaerial delta. The distributary mouth bar sands are normally 15 m thick but because of deformational processes they thicken locally to 130 m.

The characteristic depositional sequence discussed above and illustrated in Figure 3.28 indicates considerable differences in depositional rates from its basal units to the upper units. The lower sequences, the strand plain sands, clay units III, II, and I, represent accumulation rates that approximate a half a centimeter per year or less. The upper portion of the modern prodelta clays, the delta front and the distributary mouth bar, however, represent deposition associated with high accumulation rates and have been deposited within the past 600 years. Figure 3.29 illustrates accumulation rates of the mouth of South Pass, Mississsippi River, and it is based upon comparisons of numerous hydrographic maps. Since 1875 some 50 m of sediment have accumulated directly off the mouth of South

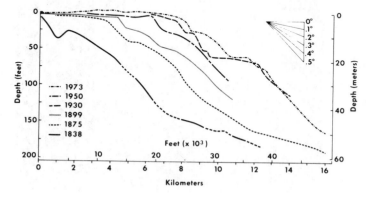

Figure 3.29. Accumulation of sediments and seaward progradation of South Pass distributary, Mississippi River, during the period 1875-1973.

Pass. Secondly, in the deeper water portions of those areas further removed from the active river mouth accumulation rates have been appreciably lower, averaging some 20 m during the period 1875 to 1975. Because of the rapid deposition of these sediments, the deposits tend to show extremely high water content; fluid pore pressures are normally above hydrostatic. As a result, the deposits are extremely unstable and are prone to a variety of subaqueous mass movements.

The rapid accumulation of a large volume of sediments over a rather restricted areal extent has resulted in abnormal weighting of the underlying Pleistocene sediments. Figure 3.30 represents an isopach map of the Recent deposits (those deposits younger than approximately 30,000 years BP) or the sediment overlying the strand plain sands of Pleistocene age. In regions adjacent (east, west, north) to the modern delta lobe, the thickness of the Recent deposits averages 50 m but in the immediate vicinity of the modern delta lobe, the sediments thicken considerably, in some cases to an excess of 250 m. The underlying strand plain sands have, therefore, been bowed or downwarped beneath the Birdfoot Delta as a result of the loading of the modern river delta sediments. The volume of the Recent sediments associated with this depocenter is some 649 km³. The implications are that sediment movement, flowage, and dewatering of the clays and silts beneath the strand plain sands have occurred as a result of modern sediment loading. The flowage has probably been occurring since early Recent deposition but with the continued accumulation of the modern delta sands it is probably accelerating and thus, in some 600 years of geologic time, a major depocenter was initiated. Figure 3.31 is a net sand distribution map of the Recent deposits based upon the same borings (several hundreds) used to construct the isopach map in Figure 3.30. The map shows contours of equal thicknesses of sands that are

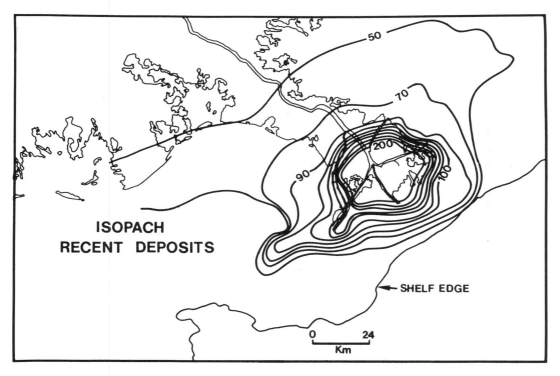

Figure 3.30. Isopach map of Recent deposits of modern Birdfoot Delta of the Mississippi River (< 35,000 years BP) Contours in meters.

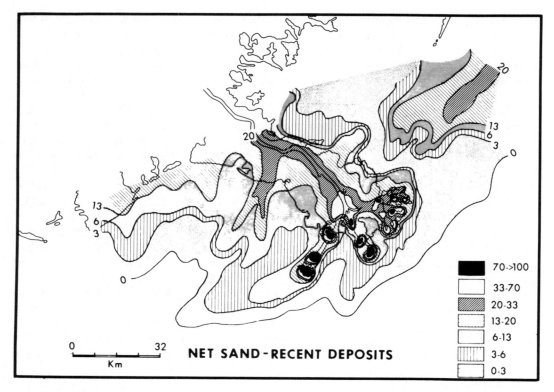

Figure 3.31. Net sand isopach of Recent deposits of modern Birdfoot Delta of the Mississippi River. (Contours in meters.)

contained within the interval from the top of the strand plain sands to the modern depositional interface. The sands associated with the distributaries are well defined. They average approximately 13 to 20 m in thickness and thin to zero in a seaward direction. Within the major distributary regions, however, are pods of exceptionally thick accumulations (greater than 100 m) of sands that are spaced repeatedly along the distributary axis. Accumulation of such thick localized sand bodies implies that the sands have been downwarped in the underlying clay causing flowage and dewatering of these sediments. This flowage results in the formations of the diapirs or mud lumps that are common at the mouth of the Mississippi River. Comparison of this map (Figure 3.31) with the isopach map shown in Figure 3.30 will show that beyond the seaward edge of these distributary sands there are still considerable thicknesses of Recent deposits. Borings indicate that this thick sequence is composed of clays and silts that have been squeezed seaward as a result of differential weighting and loading by the distributary mouth sands. The clays can thus be compared to a toothpaste tube to which a force is applied at one end and the paste squeezes out the other end. Similarly, readjustment and lateral flowage of clays caused by sand accumulation have occurred within the delta deposits and are continuing to date.

Some of these relationships described above can be seen better in a cross section of the modern Birdfoot Delta. This cross section, shown in Figure 3.32, is based on numerous cored foundation borings for offshore structures. The section runs generally around the periphery or seaward portion of the modern Birdfoot Delta. At the western edge of the cross section (Figure 3.32b), the strand plain sands are located at a depth of approximately 68 m below sea level. Overlying the Pleistocene strand plain sands are the clays associated with

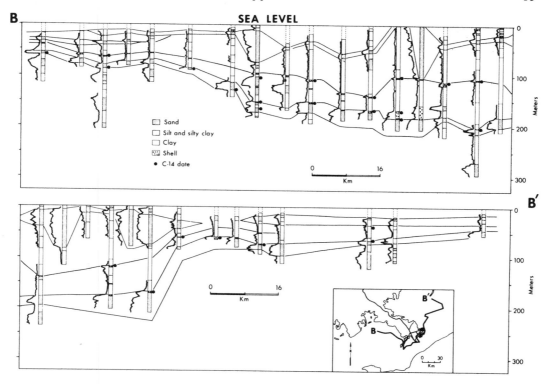

Figure 3.32. Cross section of deposits of the Birdfoot Delta. Lowermost correlative unit is a Pleistocene sand dated 35,000 years BP.

the pre-modern Mississippi River deltas. The thin sand located near the top of the section represents the regressive sands associated with the advancing Teche Delta. Moving eastward along the section, the Pleistocene or strand plain sands can be seen to bow down as the immediate vicinity of the modern Birdfoot Delta is approached. Beneath the central portion of the Birdfoot Delta, the strand plain sands can be found at a depth of 180 to 200 m beneath sea level. At a point near the easternmost edge of the modern Birdfoot Delta the strand plain sands abruptly rise in elevation to approximately 60 to 70 m below sea level. Overlying the strand plain sands in the easternmost portion of the cross section are the regressive transgressive deposits associated with the St. Bernard Delta.

The sand deposits associated with both the St. Bernard Delta and the Teche Delta average in thickness only 10 to 15 m. The facies that underlie the modern Birdfoot Delta can be seen to expand drastically. The shell horizon that separates the red clays or buff clays from the modern prodelta (Eads clays) is a highly lateral continuous marker horizon. This particular shell horizon has been dated by radio-carbon on numerous samples; nine particular dates along this line of section range in age from 13,500 to 16,600 years BP. Some 30 to 40 radio-carbon dates have been run on this horizon and thus it forms an excellent stratigraphic unit. Overlying the shell hash are the clays of the modern Birdfoot Delta. They vary in thickness considerably but as a unit they generally can be carried continuously across the entire delta. The distributary mouth bar and delta front deposits of the modern Birdfoot Delta can be seen to vary considerably in thickness. In between the major distributaries their thicknesses range from only a few meters to 10 to 12 m. As the immediate vicinity of the distributary is approached, the thickness of the sand deposits abruptly increases from a few meters to 80 to 100 m. As the thicker sands associated with the major distributaries are approached, the typical coarsening delta regressive sequence is interrupted and quite often the thickened sand bodies display rather sharp basal contacts. These sands are thickened because of differential weighting and have subsided with the underlying clays. On this scale, the subaerial deltas, which represent the crevasse splay deposits, the bay fills, the interdistributary bays, and the overbanks, form only a thin veneer that caps the major marine prodelta clay and sand deposits of the modern delta. Although considerable attention in the literature is oriented toward the subaerial deposits, in reality, they form only a thin, capping wedge over a much thicker sequence of subaqueous delta deposits. Irregular bodies of generally sharp-based sands are distributed within the prodelta. The presence of these sands in an otherwise marine clay prodelta deposit differs significantly from the typical delta profile that is commonly illustrated in the Mississippi River Delta. The explanation and process whereby this sand body is incorporated in the fine-grained clays will be discussed in the following sections.

4.1 Deformational Processes

Recent detailed marine geological investigations on the subaqueous parts of continental shelves seaward of many river deltas experiencing high depositional rates and large quantities of fine-grained sediment have revealed that contemporary recurrent subaqueous gravity-induced mass movements are common phenomena that should be considered as an integral component of normal deltaic process and marine sediment transport. Subaqueous slumping and downslope mass movement of sediments are common processes off river deltas such as the Mississippi (Coleman et al., 1980), Magdalena (Colombia) (Shepard, 1973), Orinoco (Venezuela), Surinam (Surinam), (Wells et al., 1980), Amazon (Brazil), Yukon (Alaska), Niger (Nigeria), Nile (Egypt), and Hwang Ho (China). Instabilities and mass movements of sediment in these regions generally display the following characteristics: (a) instability occurs on very low angle slopes (generally less than 2°), and (b) large quantities of sediment are transported from shallow waters offshore along well-defined transport paths and in a variety of translational slumps. The major characteristics of deltas and their offshore slopes that influence the stability of bottom sediment are as follows:

A. Rapid sedimentation results in widespread sedimentary loading of the upper delta-front slopes, especially near the mouths of passes.
B. Coarse-grained sands and silts, which compose the distributary mouth bars, differentially load the underlying and adjacent delta-front fine-grained deposits.
C. Fine-grained delta deposits, because of their rapid deposition, are generally underconsolidated, with large excess pore water pressures causing low sediment strengths.
D. Rapid biochemical degradation of organic material leads to the formation of large volumes of methane and carbon dioxide gases and in the bubble phase contribute to the generation of excess pore pressures within the sediments.
E. The offshore region experiences winter storms and hurricanes, which cause cyclic wave-induced loading of the sea floor. The cyclic loading imparts downslope stresses and contributes to pore water pressure generation.

The distribution of these characteristics as mapped from various seismic surveys, side-scan surveys, and hydrographic surveys is illustrated in Figure 3.33.

These factors in combination result in a variety of types of sediment instabilities that can be identified on low-angle offshore slopes (generally less than 0.5°). The main types of subaqueous slope instability that can be recognized in 5 to 100 m water depths are

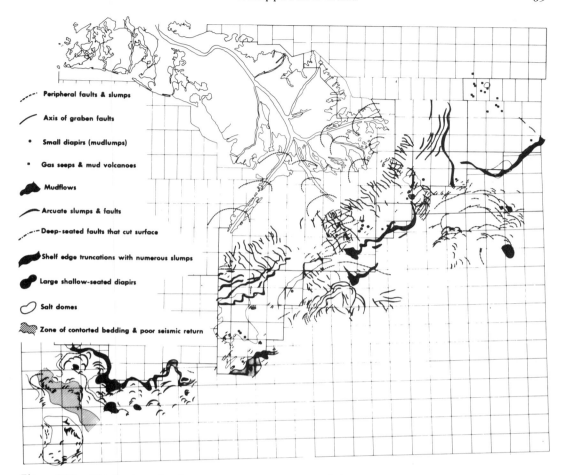

Figure 3.33. Distribution of subaqueous mass movement and faults of the continental shelf and upper continental slope off the modern Mississippi River.

illustrated schematically in Figure 3.34. Figure 3.35 shows these same features in a block diagram viewed as they would be if distributed around a single distributary. Similar features and spatial organizations can be identified in each of the three main distributary and associated bay areas of the Mississippi Delta. In the immediate vicinity of the river mouths, differential weighting by dense distributary mouth bar sands causes vertical diapiric uplift and the formation of mudlumps. In the adjacent shallow-water bays a variety of small-scale features such as collapse depressions and bottleneck slides occur. Seaward of the river mouth and on the peripheral edge of the distributary mouth bar are found a wide variety of peripheral slumps. Immediately seaward of the bar, commencing in water depths of 10 m or less, are a large number of mudflow gullies and their resulting mudflow depositional lobes. These mudflow gullies and lobes differentially weight the outer continental shelf clays and result in still other types of deformation, including a wide variety of shelf-edge arcuate slumping and contemporaneous faulting and diapirism, which are found primarily near the shelf edge and upper continental slope. Each of the main types of features will be described in detail, but it should be remembered that the categories are somewhat arbitrary, and intermediate and compound forms are also present.

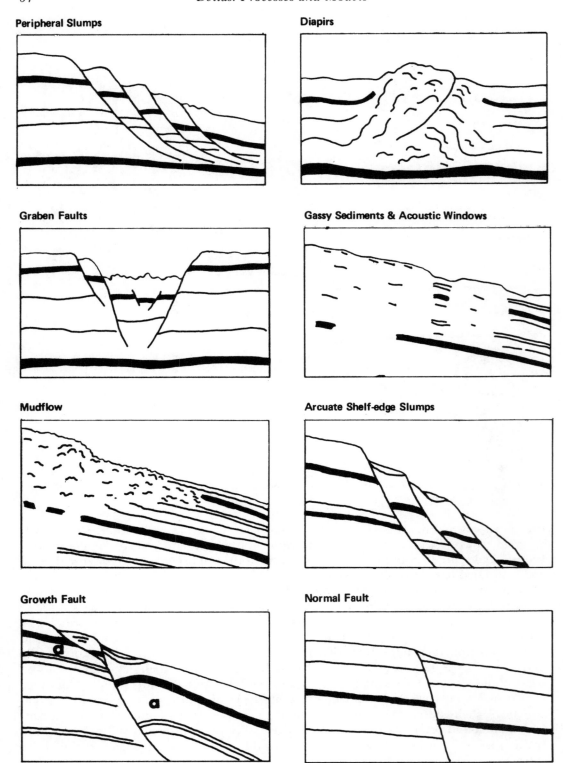

Figure 3.34. Schematic illustrations of subaqueous mass movement types.

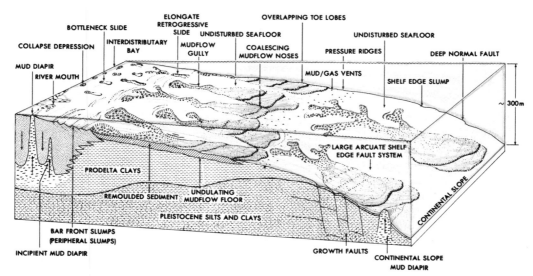

Figure 3.35. Schematic diagram illustrating the types and distributions of subaqueous sediment failures in the Mississippi River Delta.

4.2 Differential Weighting and Diapirism

Rapid deposition of localized, dense distributary mouth bar sands over less dense plastic prodelta and marine clays results in a major sediment instability. This situation leads to instability of the underlying clays, and this stress is relieved by diapiric intrusion of the clays from beneath and into the overlying sand bodies. The intrusion of the older shelf and prodelta clays into and through the bar deposits results in vertical displacement of clays by as much as 200 m and causes corresponding subsidence of the bar sands, which accumulate to thicknesses of as much as 150 m. The major mappable features that result from this differential weighting are thin spines of diapiric intruded mud, which form islands or mudlumps at the mouths of the major distributaries. Many of the diapirs do not rise above sea level, simply appearing as submarine highs in the immediate offshore region. In a single distributary-mouth pass of the Mississippi River (South Pass), covering an area of only 13 km^2, some 105 individual diapiric spines were mapped from 1876 to 1973. These diapiric spines form rapidly, and vertical displacement of more than 100 m has been documented during a period of only 20 years. Figure 3.36 is a photograph of a newly emerged diapir or mudlump at the mouth of South Pass. Although the appearance of this diapiric intrusion was quite rapid, the sediments suffer little flowage and most of the movement is accounted for by small-scale fracturing. Figure 3.37 diagrammatically illustrates the process of diapiric intrusion associated with differential weighting. Stage A depicts initial loading and compaction of the underlying prodelta and shelf deposits by progradation of the massive sands of the river-mouth bar. During stage B continued accumulation of delta-front and bar deposits accelerates lateral flowage of clays. The clays thin beneath the deltaic loading and thicken near its seaward periphery. During stage C the delta front continues to prograde seaward, and diapiric intrusion into overlying bar deposits is initiated. The load is transmitted to older and older clay units, and greater and greater amounts of vertical displacement continue. During stage D the differential loading of overlying bar sediments causes diapiric intrusion to assume a slightly asymmetrical

Figure 3.36. Photograph of mudlump island at mouth of South Pass, Mississippi River.

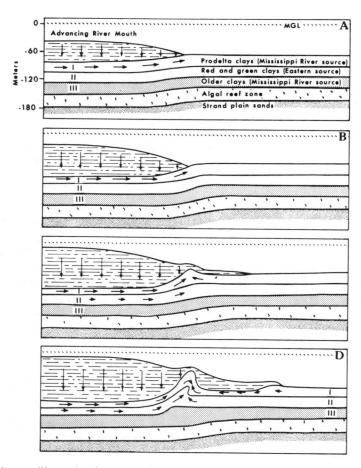

Figure 3.37. Diagram illustrating formation of mudlumps or diapirs. (Modified from Morgan, Coleman and Gagliano, 1968.)

form. Folding also affects the lower clay beds, and a submarine prominence exists on the seaward-sloping delta front of the bar deposits. In many instances high gas content is associated with diapiric intrusion of the clays, and small mud volcanoes and venting of methane gas are common. In this type of mass wasting, large volumes of older prodelta and marine clays are bowed up and intruded into the younger delta deposits. Wave reworking of the intruded marine clays results in incorporating deep-water marine micro-fauna into the shallow-water sands. Thus in some cases the shallow-water distributary-mouth-bar sands take on a marine faunal characteristic because of this process.

4.3 Collapse Depressions

These depressions occur primarily in the shallow-water areas of interdistributary bays. They are associated with slopes of 0.1° to 0.2° and with sedimentation rates that are relatively small by comparison with those of more active areas of the delta. Figure 3.38 illustrates schematically the morphology of these features. The extremely soft bottom sediments consist of recently deposited clay-rich materials and organic debris. These sediments in particular contain large amounts of methane gas, which is generated by postdepositional biogenic activity. The collapse depressions are relatively small by comparison with other mass-movement landforms in the delta. They average less than 100 m in diameter. Occasionally larger features will be found that show slightly irregular and noncircular patterns. Typically, the depressions are bounded by curved or near-circular escarpments up to 3 m high, within which the bottom is depressed and filled with irregular blocks of sediment. Side-scan-sonar records show clearly that such bowl-shaped areas, bounded by scarps, have been displaced vertically and represent distinct depressions of the sea floor. On fathometer profiles and on high-frequency seismic data, the depressed floors of the feature often show no slope and are horizontal. In addition, the floor of the feature is on many records the area where several sea-floor multiples exist, giving some indication that the sediments flooring the depression contain slightly higher densities than adjacent sediments. Similarly, side-scan-sonar records show very high return of energy from the depression floors, especially in areas between hummocky blocks.

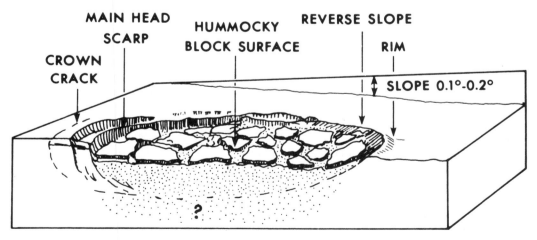

Figure 3.38. Schematic representation of the morphology of collapse depressions.

Figure 3.39 is a side-scan-sonar mosaic constructed from several track lines showing the characteristics of several collapse depressions and a small bottleneck slide. The collapse depressions are labeled A on the illustration. They range in width from 24 to 370 m and have depressed floors that range from less than 1 to 3 m below the surrounding sea floor. Irregular and hummocky topography is associated with these features. Strong reflections of the acoustic signal are often seen in the features. On slightly steeper slopes within the interdistributary regions and on slopes that approach 0.2° to 0.4° are features that are referred to as bottleneck slides. These features are similar morphologically to collapse depressions, but the boundary scarps do not form a totally closed perimeter around the instability. Rather, they show narrow openings at the downslope margin of the failure through which remolded debris is discharged over surrounding intact slopes. The bottleneck slides range in length from 150 m or so to well over 1.5 km. A typical bottleneck slide is shown in Figure 3.39c; it shows a length of 1.1 km. The source area for this feature is similar to that for collapse depressions. The bottleneck slide, however, has a large depositional lobe, labeled D in the illustration. This lobe occupies an area of approximately 6×10^4 m². Seismic data across the depositional lobe indicates that the thickness is generally less than 2 m and the feature forms a raised mound on the sea floor.

4.4 Peripheral Rotational Slides

Downslope movement of large sediment masses often begins high on the upper delta-front slope near the distributary mouths of the river. Bottom slopes immediately at the mouths of the distributaries range from 0.2° to 1.0°, but in many places major scarps show distinctive curved or curvilinear plan views. The localized scarps range in height from 3 to 7 m and contain slopes of 1° to 4°. In many cases they give the bar front a stairstepped appearance in profile view. Tensional cracks are commonly present upslope from the major scarps, and mud vents are associated with many of the scarps. The surface of the slump block normally contains extensive hummocky and irregular bottom topography and displaced clasts of sediment. The rotational nature of the downthrown block can be recognized by the reverse slope often seen in fathometer profiles and high-resolution seismic profiles. Figure 3.40 shows a schematic diagram illustrating the major morphology of these features, and Figure 3.41 shows a high-resolution seismic line run across several of these features. This type of morphology is indicative of rotational sliding over slightly curved shear planes that are concave upward, combined with subsequent translational movements in a downslope direction. Examination of high-resolution seismic data confirms that the multiple concave-upward shears tend to merge at depth into a single basal shear surface that is inclined parallel to the sediment surface. The average depth of movement of the shear plane is approximately 30 to 35 m. Cores of sediments from within the rotated blocks show laminations inclined in an upslope direction with dip angles as high as 30° by comparison with the almost horizontal attitude of the sediments nearby. Although the displacements begin as shallow rotational slumps, increasing movement downslope over the basal shear means that they become predominantly translatory in nature. This type of mass movement is responsible for moving large volumes of sediment downslope across the gentle continental shelf in front of the prograding delta. Because many of these features are confined primarily to the immediate vicinity of the distributary-mouth bars, in a large number of instances the slump blocks themselves consist of distributary-mouth-bar and distal-bar deposits that are sliding shallow-water sediment downslope to be encased in deeper water marine shales. Similar features were observed in a Cretaceous delta in Brazil by Klein et al. (1972).

Mississippi River Delta

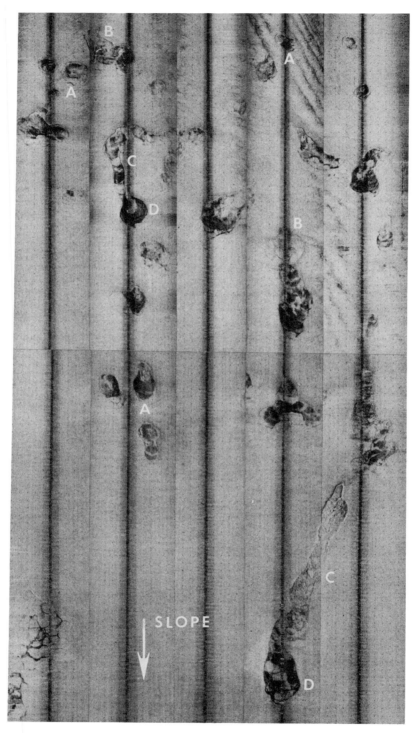

Figure 3.39. Side-scan-sonar mosaic illustrating several collapse depressions and a bottleneck slide. The grids are 25 m apart and the mosaic covers an area of 1.3 km × 2.3 km. A, collapse depressions; B, crown cracks; C, bottleneck slide; D, depositional lobe of bottleneck slide.

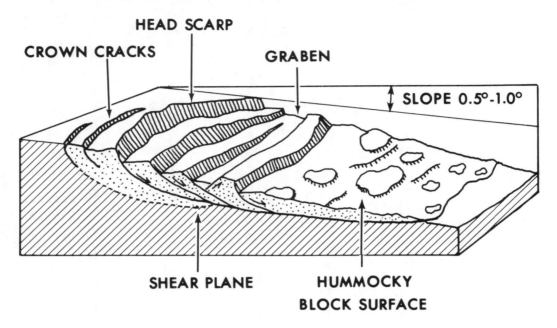

Figure 3.40. Schematic diagram illustrating the morphology of rotational peripheral slides.

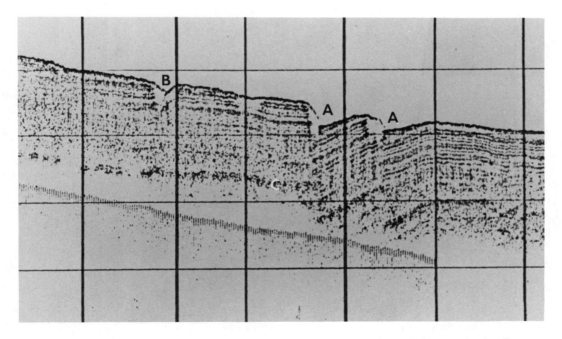

Figure 3.41. High-resolution seismic line run across several rotational peripheral slides. Navigation fixes are 152 m apart and timing lines are 10 m apart. A, rotational slides; B, newly formed rotational slide upslope; C, offset in reflection horizon.

4.5 Mass Wasting Induced by Wave Action and Degassing

The mechanisms that trigger subaqueous mass movements and slumping are complex. However, one mechanism has received considerable attention in the lower Mississippi Delta. Under extreme wave action, cyclic wave-induced bottom pressures load bottom sediments with forces not present under normal conditions (Henkle, 1970). These additional forces cause stresses within the sediments that can exceed sediment shear-strength and, as a result, bottom sediments assume new conditions of equilibrium. Figure 3.35 illustrates regions also characterized by locally closed depressions. These closed depressions are found throughout the delta front in water depths of from only a few meters to in excess of 120 m. In many instances bottom slopes in these areas characterized by these features range from about 0.6° (1%) to 0.3° (.5%). The steepness of such slopes is far less than critical for these sediments (about 2° to 5°), and yet slump-like mass movements are evident. Features similar to those illustrated in Figure 3.38 have been documented by Bea and Arnold (1973) and Sterling and Strohbeck (1973) in 105 m of water in Block 70, South Pass, Mississippi River Delta. Movement of features in this block was associated with wave action during the passage of Hurricane Camille in 1969, during which time Shell Oil Company lost one platform and another was moved laterally in this region. The sediment movement extended to a depth of 21 m; this determination was made from the depth at which the rig piling failed. Relationships of movements in low-strength sediments under cyclic wave loads have long been known and documented. Laboratory observations by Doyle (1973) indicate that a wave-like periodic motion of the bottom, coupled to a surface wave, can occur. This bottom wave is analagous to an internal wave at the mudline. It may show a height of several feet and produce orbital motions as deep as 17 m (Bea and Arnold, 1973). Direct field evidence for similar phenomena has been documented by Suhayda et al. (1976) and Tubman and Suhayda (1976). Because of the vast literature on this subject, the details of the mechanism and the coupling of surface waves with bottom motion will not be given here.

One aspect of this type of deformation generally associated with wave action has received only minor scientific attention. The fine-grained sediments surrounding the Mississippi River Delta contain abundant sedimentary gases. The high gas content may produce a pronounced effect on sediment stability directly by entrapment of gas bubbles in the unconsolidated sediment or indirectly by the upward migration of these gases as a result of bottom pressure perturbances. Dissolved gas concentrations (methane) in the shallow lower delta sediments range from 0.047 to 150 ml/l. This microbial CH_4 is produced within anoxic sediments, and high CH_4 concentrations are usually associated with sedimentary environments rich in organic carbon. The decay of organic matter is often accompanied by bacterial SO_2 reduction. During this process CO_2 is produced; the CO_2 then combines with sedimentary H_2 (from organic matter) to produce methane according to the following reaction: $CO_2 + 4H_2 = CH_4 + 2H_2O$. In marine sediments there may exist competition for hydrogen between sulfate-reducing bacteria and methane-producing bacteria. Because CH_4 production begins as sulfate reduction diminishes, areas of low sulfate and high dissolved organic matter in the sediment pore water would have the chemical capability of producing methane by this reaction. In the sediments off the Mississippi Delta the high seasonal discharge of fresh water lowers the sulfate concentration in nearshore waters. These waters and the fine-grained sediment contain high concentrations of organic material. These combinations would allow the dissolved CO_2 to react with available molecular hydrogen to form methane. Thus conditions are ideal in the Mississippi River sediments for the production of abundant sedimentary gases. These processes have been described adequately by Whelan et al. (1975, 1976).

The production of large volumes of in situ gas alone may be enough to alter the stability of the sediments so that they eventually fail and flow seaward. However, cyclic perturbation of bottom pressure induced by the passage of storm waves could cause the sediments to lose their stability even more quickly than those sediments containing less gas concentration. Bottom sediments may be induced to move as a fluid-like mass because of the presence of these gases, a fact of significant importance in engineering considerations. With the advent of a hurricane wave, extreme changes occur in the gas-rich bottom sediments; the ambient pressure and the pressure that the gas is subject to now varies with the wave frequency. The bottom pressure change causes the gas in the sediments to alternately contract and expand. As the storm waves increase in height and period, the induced bottom pressures increase and may cause such a large expansion of gas that sediment grains would separate, releasing large amounts of gas at one time. This activity could be the trigger for the initially loosely packed sediments (which have been separated even more by gas expansion) to collapse and become fluidized because bottom sediment shear strengths would vanish. The sediments would tend to collapse by degassing and dewatering and move down any sloping gradients owing to the action of gravity. Immediately after the hurricane, the bottom sediments would be more densely packed; porosity and gas content would be expected to be lower. If this assumption is true, wave action in combination with gas content plays a significant role in subaqueous mass movement along the fronts of the Mississippi River Delta.

High gas content in the sediment can be associated normally with the loss of acoustic return, and in many cases no subbottom reflectors are present in the regions displaying a high concentration of sedimentary gases. In many parts of the Mississippi River Delta poor seismic returns are commonplace. Within this acoustically transparent region, however, are found numerous acoustic windows through which seismic reflections can be seen. These windows are commonly associated with closed depressions as mapped from side-scan-sonar and hydrographic surveys, such as those illustrated in Figure 3.39.

4.6 Mudflow Gullies and Depositional Lobes

Extending radially seaward from each of the distributaries in water depths of 7 to 100 m are major elongate systems of sediment instabilities referred to as delta-front gullies or mudflow gullies. The features were first described from hydrographic maps by Shepard (1955), and an illustration of a topographic configuration is given in Figure 3.42. The numerous topographic gullies that scar the delta platform appear on the hydrographic map. Side-scan-sonar records and high-resolution seismic data show that these valleys or gullies emerge from within an extremely disturbed area of slump topography high on the delta. Each gully shows a clearly recognizable area of rotational instability or shear slumps at its upslope margin. This feature is the most common type of sediment instability fronting the Mississippi River Delta. Each gully possesses a long, sinuous, narrow chute or channel that links a depressed hummocky source area on the upslope margin to composite overlapping depositional mudflow lobes on the seaward end. Figure 3.43 illustrates schematically the major morphological characteristics of these features. The instability is bounded on its upslope side by a bowl-shaped depression that serves as the source area. Often, multiple head scarps and crown cracks can be seen on the side-scan-sonar records, indicating upslope retrogression. Within the bowl-shaped depression, hummocky, irregular, distinctive blocks of various sizes and arrangements can be discerned. Downslope from the bowl-shaped source area is an essentially elongate chute. These chutes or gullies are bounded by very sharp linear escarpments that are arranged parallel or subparallel to one

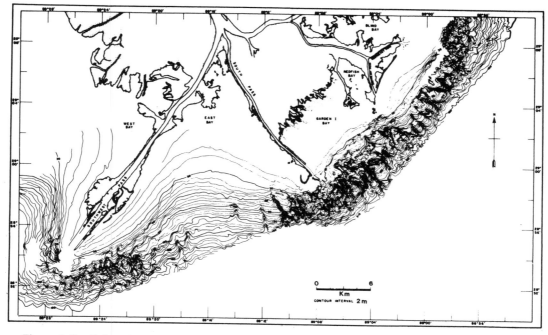

Figure 3.42. Submarine topography of upper slope off Mississippi Delta (after Shepard, 1955.)

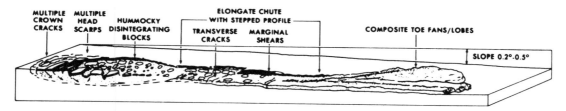

Figure 3.43. Schematic diagram illustrating the morphology of mudflow gullies and depositional mudflow lobes.

another. The area enclosed by the scarps is downthrown and is composed of irregular, chaotic topography of blocks of debris of varying sizes. Commonly the blocks within the chute area are smaller toward the central axis of the gully. The gully floors are 3 m to as much as 20 m below the adjacent intact bottom. The slopes along the sides of the gullies range from less than 1° to as high as 19°. Most of the gullies extend downslope approximately at right angles to the depth contours and may be more than 7 to 10 km long.

In plan view, the features are rarely straight, and quite commonly are markedly sinuous and show alternating narrow constrictions or chutes and wider bulbous sections. Figure 3.44 is a map showing the mudflow gullies in a small region located to the southwest of the mouth of South Pass. It shows in plan view various aspects of the radial gullies or mudflows as they have been mapped utilizing overlapping side-scan-sonar data. Four major gullies can be discerned, the longest of which continues for distances in excess of 4 km and continues off the map for another 4 or 5 km downslope, ending in approximately 90 m

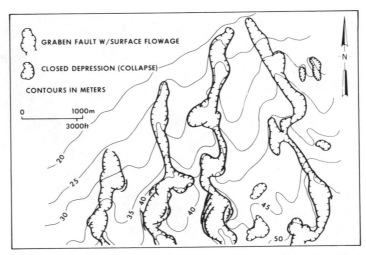

Figure 3.44. Surface geologic map, based on side-scan-sonar surveys, showing radial grabens in vicinity of South Pass.

water depth. Figure 3.45 is a side-scan-sonar mosaic constructed from lines run across a zone of landslide gullies. The area covered by the mosaic is 1.2 x 1.8 km. Three major elongate gully systems are shown, beginning with blocky source areas (A) bounded by scarps. The source area geometry is very irregular; considerable differences exist in block size and orientation. One gully has retrogressed upslope and incised an adjacent well-established gully (B). The narrow gullies are relatively deeply incised (C), and evidence of side-wall instability is indicated by small slumps along the gully margin (D) and by the alternations of bulbous source areas and narrow chutes. The widths of the individual gullies range from 18 to 150 m at the narrow points to 370 to 550 m at the widest. The floors of the gullies are characterized by large erratic blocks of different sizes (E) found in complexly fractured remolded debris. At the down-slope ends of the gullies extreme widths of as much as 2 km can be found.

The side walls of the mudflow gullies are subject to instability; this slumping can produce contrast in forms and is probably responsible for localized widening along an individual gully system. Figure 3.46 shows a high-resolution seismic line run at right angles to the axis of a narrow gully. Several rotational slides (A) on both sides of the channel can be discerned. After the blocks slump down into the gully, they are carried further downslope during the next episodic movement of the debris in the gully. The offset of reflection horizons and the stairstepped topography expresses this instability. The formation of elongate chutes of this type is very similar to the morphology associated with subaerial debris flows and some types of mudflows. The chutes or channels emanate generally from upslope slump zones and constitute transport conduits for disturbed and remolded sediments, together with displaced blocks of various sizes. The mechanism of transport is probably characterized as slurry flow, which can be a type of plug flow in which rigid plugs move over and within a zone of liquefied mud. The presence of partially disintegrated rafted blocks suggests laminar or plug flow rather than turbulent flow. Often cores taken within the gully floors indicate good preservation of finely laminated sands and silts, indicating transport as a series of blocks. The only evidence of transported debris downslope is displaced fauna within the debris and extremely high depositional dip angles within the deposits. The shear planes beneath the transport gullies are very thin, and extremely closely spaced corings are required to permit recognition of a shear plane.

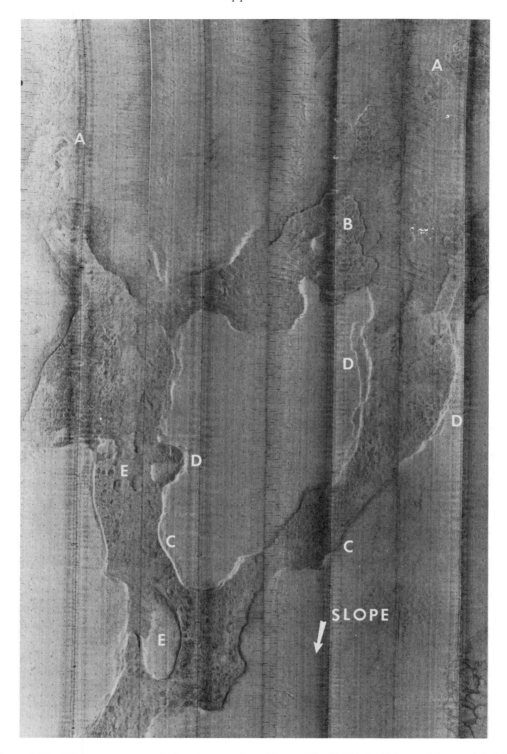

Figure 3.45. Side-scan-sonar mosaic showing several mudflow gullies. Grid is 25 × 24 m, and the mosaic is 1.2 × 1.8 km. Water depths are approximately 21 m at the top of the figure and 34 m at the lower end. A, source area; B, retrogressive gully; C, narrow incised gully; D, side-wall instability slides; E, large erratic blocks in gully floor.

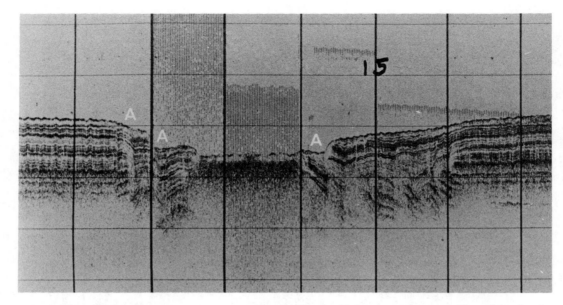

Figure 3.46. High-resolution seismic line run across a mudflow gully that shows numerous side-wall rotational slides (A). Navigation fixes are 152 m apart and timing lines are 10 m apart.

At the seaward or downslope ends of the mudflow gullies extensive areas of irregular bottom topography are composed of discharged blocky, disturbed debris. In plan view this discharged debris is arranged into widespread overlapping lobes or fans. This morphological feature is illustrated schematically in Figure 3.43. Each lobe is composed of two morphological features: an almost flat or gently inclined surface (less than 0.5°) and an abrupt distal scarp representing the downslope nose of the displaced debris. The seaward scarps range in height from only a few meters to more than 25 m and have slopes as great as 7° to 10°. In plan view the scarps are generally curved, and adjacent lobes are separated from one another by major reentrants. Because of the large number of mudflow gullies that front the present delta, the displaced debris from adjacent gullies may coalesce, providing an almost continuous sinuous frontal scarp that extends peripheral to the modern bird-foot delta. Detailed mapping, however, shows that the depositional areas are composed of multiple overlapping lobes, each having its own distinctive seaward nose, and are due to episodic discharge from the gullies farther upslope. The more recent the emplacement of a lobe, the more irregular and blocky the surface topography; in older depositional lobes the topography is commonly characterized by small-scale pressure ridges arranged as sinuous parallel ridges and hollows and in places contain many small mud volcanoes and gas vents produced by localized sedimentary loading.

Figure 3.47 is a side-scan-sonar mosaic (1.5 x 2.1 km) of a depositional mudflow lobe emanating from an upslope mudflow gully off South Pass, Mississippi River Delta. At least three overlapping lobes (A, B, and C) comprise this feature. The discharge debris consists of extremely erratic large blocks, most of which are about 30 m or so in diameter. Larger blocks (D) are often incorporated in the depositional lobes, and may be 150 to 300 m in diameter. The average thickness of these individual lobes is 10 to 15 m; however, high variation exists from one part of the delta to another. Figures 3.48 and 3.49 are high-resolution seismic lines run across depositional mudflow lobes. The seismic line illustrated in Figure 3.48 shows a major scarp on the sea floor (vertical scale is in seconds, where 100 ms equals 76 m), which is the seaward terminus of the mudflow in approximately 86 m of water. Seaward of this scarp, the mudflow continues as a thin amorphous unit seaward for a distance of 10 km (beyond

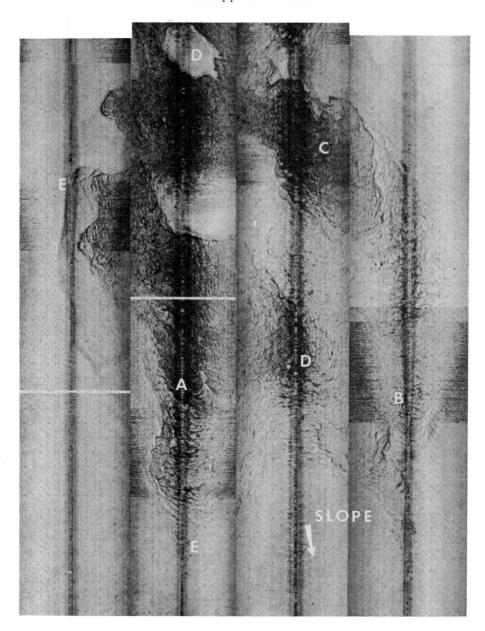

Figure 3.47. Side-scan-sonar mosaic showing multiple overlapping mudflow depositional lobes. The mosaic covers an area 1.5 × 2.2 km of sea floor. The grid marks are 25 m apart. A, B, C, mudflow lobes; D, erratic blocks; E, pressure ridges.

the left-hand margin of the diagram). Cores within this mudflow tend to indicate a high gas content, and it is likely that this gas accounts for the lack of seismic reflections within the mudflow.

In Figure 3.49, the vertical scale is in seconds (100 ms equals 76 m), and the length of the line is approximately 10 km. This mudflow has moved beyond the shelf and is presently located at approximately the shelf upper slope break, which occurs at approximately 200 m water depth. This particular lobe is composed of multiple overlapping lobes; however, in this diagram the entire mass is fairly amorphous and is approximately 50 m thick.

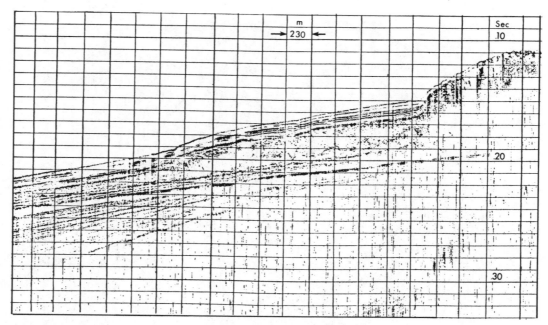

Figure 3.48. High resolution seismic record (6.5 kHz) run across mudflow.

Figure 3.50 illustrates the distribution of these mudflow gullies, a few collapse depressions, and peripheral slumps off a part of the Mississippi River Delta. The particular area is seaward of Southwest Pass, and it shows numerous mudflow gullies that crease the delta front, with their depositional mudflow lobes spread out across the shelf in front of the delta. Studies of the entire delta show that this is a major transport mechanism of moving shallow-water sediment across the shelf into deeper offshore waters. Approximately 40% of the sediment that flows down the major distributaries of the Mississippi is involved in mass movement and results in the transport of large quantities of shallow-water sediment across the shelf to deeper water.

4.7 Shelf-Edge Slumps and Contemporaneous Faults

Another major type of sediment instability that has significant geological importance is the arcuate rotational slumps and growth or contemporaneous faults that commonly occur on the outer continental shelf in front of the prograding deltaic system. Large arcuate families of shelf-edge slumps and deeper seated contemporaneous faults tend to be presently active along the peripheral margins of the delta fronts. The distribution of some of these features off the mouth of South Pass is illustrated in Figure 3.51. These large-scale features cut the modern sediment surface, often forming localized scarps on the sea floor, and these surface scarps provide localized areas for accumulation of downslope mass-moved shallow-water sediment. The contemporaneous slumps tend to show a concave plan-view pattern (Figure 3.52). The shear planes are concave upward and tend to merge into bedding planes with depth. The shelf-edge slumps and slump faults tend to give a stair-stepped appearance to the edge of the continental shelf and are highly reminiscent of the rotational peripheral slumps higher up on the continental shelf, near the mouths of the modern distributaries. However, these features are generally on a much larger scale and cut a sediment column ranging from 50 to 250 m in thickness. Lateral continuities of individual slump scars range

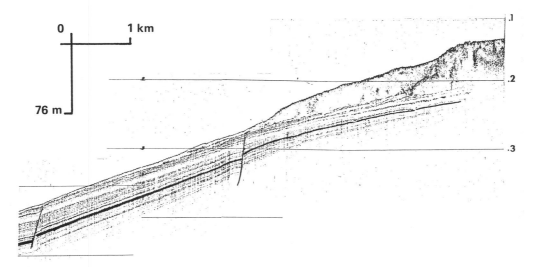

Figure 3.49. High-resolution seismic line run across a depositional mudflow lobe.

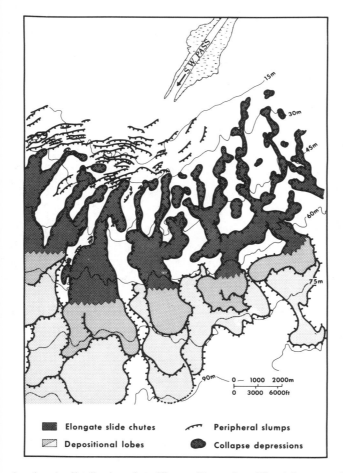

Figure 3.50. Map showing the distribution of mudflow gullies and mudflow lobes around a portion of the delta off Southwest Pass.

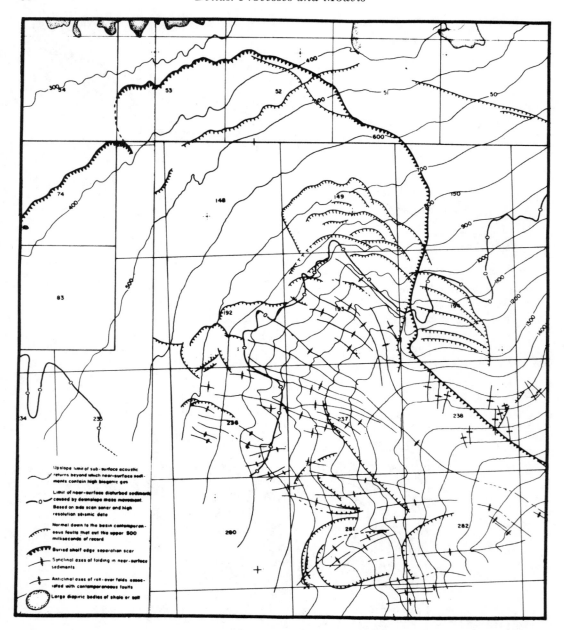

Figure 3.51. Map showing the distribution of contemporaneous faults off South Pass, Mississippi River Delta. These faults occur approximately 23 km seaward of the river mouth and in water depths that approximate 250 to 300 m.

from a few kilometres to as much as 8 to 10 km, and the scarps on the sea floor produced by this slumping process are as high as 30 m. Contemporaneous slump faults show recent movement along the shear planes and have formed a scarp on the sea floor. Movement along the fault is contemporaneous with deposition, and hence larger amounts of sediment normally accumulate on the downthrown side of the fault. This particular series of faults extends from the present sediment surface to depths beyond the bottom of the seismic record. Lower frequency sparker data run simultaneously indicate that the fault extends 700 to 800 m below the sea bottom before merging into a bedding plane fault. Offsets in the uppermost units are generally on the order of 5 to 10 m, whereas at depth offsets of marker beds approach 70 to 80 m. The sediment units show as increased thickness on the downthrown side of the fault and the small rollover anticline or reverse drag characteristic of this type of fault. The amorphous zone upslope represents a surface mudflow that has progressed to and slightly beyond the limits of the fault. As the surface mudflow crosses the fault zone, it becomes thicker, and it is highly possible that increased thickness on the downthrown sides of these faults is the result of the movement of surface mudflows across the fault zone. As the fault zone is blanketed by a large mass of rapidly introduced mass-movement sediment, surface scarps on the sea floor are thus eliminated. Continued movement along the fault, however, will cause a new scarp to form and, given enough time, another mudflow will then move across the feature, adding increased amounts of sediment to the downthrown side. This type of interaction between surface mudflow movements and contemporaneous faults quite possibly may play a large role in maintaining the continuing movement along the fault planes.

A feature commonly associated with the growth faults, and of extreme importance to petroleum trapping, is the association of rollover or reverse drag on the downthrown side of the growth fault (Figure 3.52). Such features are common on the contemporaneous faults that are presently active in the delta. The rollover anticline tends to form soon after deposition of the sediment on the downthrown side and does not require a considerable amount of overburden and weighting in order to form. The mass-moved material that flows downslope from higher levels on the delta front (sands, silts, and clays) has high water and gas contents. It is speculated that, as the sediment accumulates slightly more thickly on the downthrown side of the fault, early degassing and dewatering take place associated with movement along the fault. Pore waters and pore gases are permitted to escape upward in the zone of movement associated with the fault, thus decreasing the volume of the sediment and allowing an early change in density to take place nearly contemporaneously with development of the fault. As greater and greater amounts of sediment are added and overburden pressures become increasingly greater, this feature is amplified and becomes more pronounced with time and depth.

Figure 3.53 is a cored boring through a slump block off Southwest Pass, Mississippi River Delta. Each core was 9 cm in diameter and 1.8 m long. Core 7 is a marine shale beneath the slump block. The sediment in it is highly distorted and shows fractures and other types of disturbance. Cores 2 to 6 represent the slump block itself. The lower section of the slump block, cores 5 and 6, is undoubtedly distal-bar deposits, the zone from which the shear plane developed. The dips are of extremely high angles. The cores were unoriented, and thus the precise dip direction is not known. Cores 2, 3, and 4 are primarily sand deposits. These sands contain sedimentary structures that are very reminiscent of the distributary-mouth bar. The abundance and variety of distorted structures and some of the fracturing within these deposits are clearly shown in Figure 3.53. Core 1 is a normal marine sediment, deposited in the mid- to outer-shelf depth range. It tends to cap the sand deposit and is probably not representative of the slumping process itself. It is highly possible that this represents fine-grained marine clays that capped the slump deposit once it had occurred.

4.8 Summary of Deformational Features

Various types of subaqueous mass movement processes have been identified in the Mississippi River Delta region and, although complex in nature, do have a systematic distribution pattern, as shown in Figure 3.35. Rotational peripheral slumps, diapiric intrusion, and mudflow gullies are more commonly associated with the major distributar-

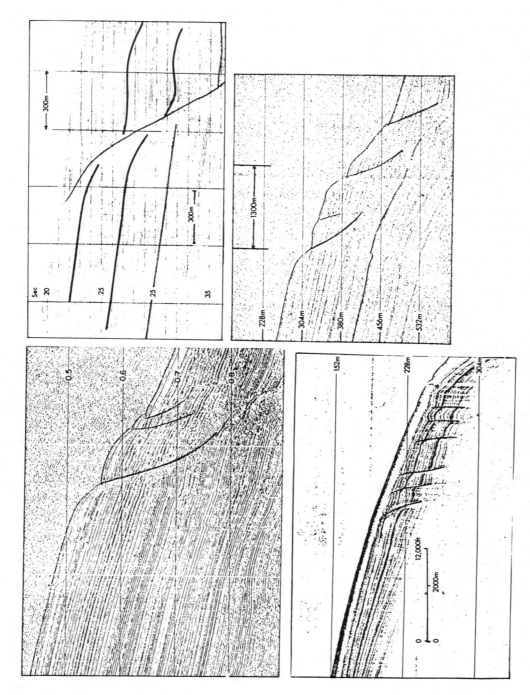

Figure 3.52. High-resolution seismic lines run across a variety of types of shelf-edge slumps and contemporaneous slump faults.

Figure 3.53. Cored boring through a slump deposit off Southwest Pass, Mississippi River Delta. Scale is in feet and tenths of feet.

ies of the delta, but occasionally do occur in regions removed from the immediate river mouths. These types of mass movement result in considerable rearrangement of facies along the delta front and feed large volumes of sediment to deeper water depths surrounding major delta lobes. Localized slumping and bed fluidization, triggered by bottom-pressure fluctuations associated with large surface waves, tend to form in those regions of the shallower waters across the entire delta front to water depths on the order of 100 to 150 m. Sedimentary gases generated primarily by bacterial degradation of organic material play a significant role in the bottom mass movement of this type. Such early changes in the density of the deposited material can affect significantly early diagenetic changes and fluid movement and migration through the delta-front sediments. One of the major types of subaqueous mass movement consists of large surface mudflows, which deliver considerable volumes of sediment from higher portions of the delta to and across the outer continental shelf and upper continental slope. This mechanism plays an important role in shelf sediment transport and under certain conditions can result in accumulations of shallow-water deltaic sands encased in deeper marine clays far removed from the original site of deposition. A variety of faults, especially the arcuate shelf-edge faults and slumps and the contemporaneous faults, tend to form peripheral to the deforming load of the major deltaic depocenter. Thus, the pattern of major faulting within a depocenter is generated early in the history of development in the deltaic facies. Recognition of these features sheds new light on the facies relationships generally considered in deltaic sediments. These facies are summarized in the cross section illustrated in Figure 3.32.

Although these features are common to the Mississippi Delta Front and Prodelta zone, they have rarely been observed in ancient counterparts. Two possible cases have been reported from the Cretaceous Parkman Sandstone of Wyoming (Hubert et al., 1972) and the Cretaceous Pitanga Member of the Candeas Formation of the Reconcavo Basin of Brazil (Klein et al., 1972). In the Parkman, thin rollover folds are common. Within the Pitanga Member, gullies on delta fronts are preserved and filled with sandstones showing obliteration of porosity, slump folds and faults and subaqueous mudflow and debris-flow conglomerates and conglomeratic sandstones. These features were observed both in outcrop and in the subsurface.

4 | VARIABILITY OF MODERN DELTAS

The processes and factors described previously exert significant control on the geometry, genesis and distribution of deltaic facies. Systematic comparisons of a large number of deltas have indicated that there are distinct groupings among modern-day deltas; details of comparisons of numerous modern-day deltas have been published by Coleman and Wright (1971), Wright, Coleman and Erickson (1974), and Coleman and Wright (1975). Their studies consisted of systematically acquiring, tabulating, and generating similar information from major world deltas, so that statistical comparisons could be made which would indicate the processes that exert significant control in delta facies. Although 55 deltas were initially chosen for the comparisons, only 34 had sufficient data for statistical comparisons; these deltas are indicated in Figure 1.1, while Table 1.1 lists the major characteristics. The river systems studied represent a wide variety of environmental settings. They span all climatic zones, range in size from extremely small to large river systems, debouch into receiving basins that range from large oceans to enclosed basins, and cover a wide variety of tectonic settings. The variety of these process settings is representative of most existing global settings and, it is believed, covers many conditions that existed in past geologic times.

The approach utilized is illustrated in Figure 4.1. The initial step of the project was to define rigidly the various component parts of a river system and the parameters to be measured and tabulated. The definitions were based on experience, published literature, and discussions with other scientists. Following this step, reconnaissance trips to deltas in various parts of the world were conducted to ascertain the validity of the definitions and to modify them if necessary. Following this field work, a compilation and evaluation was completed of the existing data on each delta. Much of the raw data were derived from published literature, maps, aerial photographs and various government agencies and consulting engineering firms. It was during this stage that the large number of variables were decreased and selected because of standardization of data. Standardized techniques for measuring raw data were also developed. One of the major problems encountered was optimization of scales for maps and photographs so that comparable parameters could be obtained from these data. The various data for each delta were then compiled and utilized to generate specific process and form information. At the outset of the program, emphasis was placed on obtaining quantitative data and, as a result, various computer programs were developed for quantifying various parameters. An example consisted of quantifying the wave energy along the world's major deltas (Coleman and Wright, 1971; Wright and

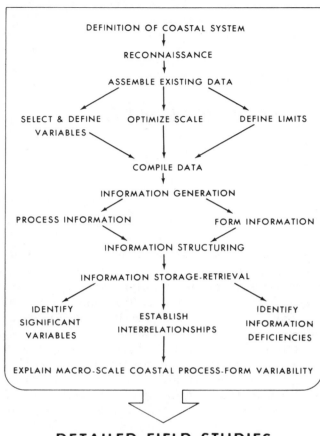

Figure 4.1 Flow chart showing approach to comparison study of deltas. (Republished by permission of the Houston Geological Society; Coleman and Wright, 1975.)

Coleman, 1973). Quantification of these selected parameters is extremely important; in much of the existing published literature, quantification was not attempted and has resulted in numerous incorrect evaluations of certain parameters. This limitation is especially true in the wave power/wave energy aspect of deltas. Several authors have simply referred to high wave energy or low wave energy deltas; often the published literature is contradictory in nature. Numerous other information-generating techniques were developed during this stage of the project. The resulting information and data were then structured, a format was developed, and all data were stored in a computer program for efficient handling and retrieval. Once the data bank was completed, various statistical techniques were utilized to compare and ascertain similarity and differences between deltas. The small sample size of deltas in relationship to the large number of variables available precluded use of differential analysis such as multiple regression; instead the approach was to examine the affinities and differences between deltas in terms of several variables using a combination of multivariate analyses, including cluster, factor and discriminant analyses. The conclusions of this aspect have been presented by Wright, Coleman and Erickson (1974). The final products were determination of the process-form variability of modern world deltas and attempt at identification of delta models. Table 1.1 indicates a summary of the various parameters that were gathered for the 34 deltas utilized in the project.

In order to present the data in a geological framework, 6 modern world deltas have been chosen to illustrate the relationship between the process setting and the resulting sedimentary sequence. The following deltas will be described:

1. Mississippi Delta (U.S.A.)—low wave energy, low tide range, low offshore slope, low littoral currents, large fine-grained suspended load, unstable receiving basin, and a temperate climate;
2. Klang Delta (Malaysia)—low wave energy, high tide range, high littoral currents, narrow seaway receiving basin, and a tropical climate;
3. Ord Delta (Australia)—low wave energy, extreme tide range, low littoral currents, restricted receiving basin, and an arid climate;
4. Burdekin Delta (Australia)—intermediate wave energy, high tide range, low littoral drift, stable receiving basin and a semi-arid climate;
5. Sao Francisco Delta (Brazil)—high persistent wave energy, intermediate tide range, low littoral currents, steep offshore slope, and dry tropical climate;
6. Senegal Delta (Africa)—extreme wave energy, intermediate tide range, high littoral currents, steep offshore slopes, and an arid tropical climate.

Significant characteristics of these deltas are summarized in Table 4.1. A general description of each delta will be presented and then a composite vertical stratigraphic column is discussed. This approach of utilizing a vertical sequence has been chosen as it allows a description of each of the major lithofacies and environments of deposition present in each of the deltas. The composite stratigraphic column, then, represents a complete sequence of all of the environments as they would exist as seen in a vertical section or as mapped in an outcrop area. The probability of encountering such a complete sequence is rare and some indication of the most common sequence is given below each column. In many instances lateral relationships are unknown and hence little reference will be made to this aspect.

Table 4.1. Significant Process Parameters of the Delta Examples

Delta	Climate	Discharge (m^2/sec)	Sediment	Wave Power (ergs/sec x 10^6)	Tide (m)	Nearshore Currents	Shelf Slope (%)	Tectonics
Mississippi	Humid subtropic	15,631	High suspended	0.034	0.43	Low	7.0	Rapid Subsidence
Klang	Humid tropic	1,100	High suspended	0.218	4.2	Extremely high	4.1	Low Subsidence
Ord	Dry subtropic	166	Bed load and suspended load equal	1.06	5.8	Low	3.9	Stable
Burdekin	Dry tropic	476	High bed load	6.41	2.2	Low	9.2	Stable
Sao Francisco	Dry tropic	3,420	High suspended	30.42	1.9	Low	11.2	Stable
Senegal	Arid tropic	867	High suspended	112.42	1.2	Extremely high	17.0	Low Subsidence

1. Mississippi River Delta

The Mississippi, the largest river system in North America, drains an area of 3.3×10^5 km². Average discharge at the delta apex is 15,360 cumecs (cu m/sec). Sediment discharge is estimated at 2.4×10^{11} kg/year; the sediments consist predominantly of clay, silt, and fine sand. The details of the various environments of deposition have been presented in the previous chapter and will not be repeated in great detail in this discussion. Figure 4.2 shows a general distribution of depositional environments of the Mississippi River Delta. The delta plain has a total area of 28,568 km² of which 23,900 km² is subaerial. Marshes and open bays form the bulk of the subaerial delta deposits. The warm humid climate favors in situ accumulation of peats in the marsh areas and the bays are characterized by highly burrowed silts and clays. Shoreline wave power is extremely low, averaging only 0.034×10^7 ergs/sec. Riverine processes dominate and the river mouths protrude out into the receiving basin as long finger-like configurations. The coastline, when viewed from the vertical, displays an extremely irregular shoreline. Brackish water bays, broad coastal marshes, and swamps form the most distinctive landforms between the long finger-like distributaries. Beach ridges and other marine reworked deltaic sand bodies are relatively rare and exist only in the immediate vicinity of the abandoned river mouths. When present, these deposits are thin and display little or no lateral continuity. The bulk of the subaerial deposits consists of interdistributary bay deposits and crevassing or overbank splays which tend to fill the brackish water depressions. The smaller overbank splays emanate from the main river channel and splay out into adjacent interdistributary bays as small lenses of sand. The larger crevasses or subdeltas break or scour across the natural levee of the river channel and form deposits with an areal extent on the order of 160 km²; they tend to form small deltaic lobes with thicknesses of 6 to 15 m. These subdeltas are virtually the only sands that occur within the subaerial delta deposits.

The Mississippi River carries a high suspended sediment load and fine-grained deposits are spread laterally over large areas seaward of the river mouths. The wide lateral dissemination of this fine-grained sediment builds an unstable platform across which the distributaries must prograde. As a result, differential compaction and formation of thick distributary sands, diapiric clay intrusions (mudlumps), downslope mass movement and slumping, and a variety of faulting are characteristic of the subaqueous delta deposits. These processes have been described in Chapter 3. The major sand bodies of the subaqueous delta consist of the distributary mouth bar deposit which thickens and thins along the axis of the distributary in response to differential subsidence. Once the active channel has been abandoned, channel filling is a process in which fine-grained, organic rich clays and silts accumulate in the channel and form virtually a fine-grained channel fill deposit. In some instances sand bodies are found lower down on the delta slope. These sand bodies result from a variety of subaqueous mass movement processes and these shallow water sands are generally encased in marine, prodelta, or shelf clays.

The composite stratigraphic section of the Mississippi River Delta is illustrated in Figure 4.3. There, the lowermost units consist primarily of fine-grained clays and silts that represent prodelta and shelf deposits. These deposits generally tend to show wide lateral continuity. Occasionally within these fine-grained deposits are found thin sand bodies (Figure 4.3: Unit 2) representing slump blocks that were moved downslope and subsequently encased in marine shelves; the sands normally contain a shallow water fauna. The lateral continuity of these sands is low, but borings within the lower delta generally encounter several sands stacked vertically within the clay rich deposits indicating the importance of this process in this type of delta. One of the most characteristic features of these sands are the high dip angles, 20° to 50°, and when present contain shallow water fauna. In some instances the top of the sands display high burrowing by organisms. Unit 4 (Figure 4.3) represents the delta front deposits of the prograding deltaic sequence, and although characterized primarily as alternating sands, silts, and clay, occasionally thin well-sorted sand deposits are found within this unit. Lateral continuity of the sands, however, is low. The major sand body present in a Mississippi River type delta consists of the distributary mouth bar deposits (Figure 4.3: Unit 5). The distributary mouth bar sands normally accumulate up to thicknesses of 12 to 20 m, but in localized regions, abrupt thickening of this sand body can be found and thicknesses of up to 100 to 135 m are common. The thickened pods, when present, normally display a rather sharp base and can be easily misinterpreted as a sand-fill channel deposit. Often associated with the distributary mouth bar will be a thin, well-sorted sand that caps this unit. This sand unit (Figure 4.3: Unit 6) represents wave reworking of the uppermost portion of the distributary-mouth bar deposits and concentration of the coarser sand into a small beach deposit. The beach deposits show little lateral continuity and generally do not attain any great thickness. Because of the large amount of organic debris carried downstream by the Mississippi River, transported organic lenses, up to several meters thick, are found in this unit and in the uppermost portion of the distributary-mouth bar. These organic deposits would form coal stringers in sandstone and would be characterized by a lack of rooting on the basal portion of the coal deposit.

Unit 7 (Figure 4.3) consists of overbank splay deposits or natural levees. This deposit is normally composed of interbedded silts, sands, and silty clays commonly characterized by intense plant-rooting structures. In some instances, early diagenesis of the deposits occurs and the sediments are cemented with siderite and calcium carbonate. Interdistributary bar deposits (Figure 4.3: Unit 8) display varying thicknesses and a wide variety of lithofacies. Often, the interdistributary bays consist primarily of highly bioturbated clays and silts, whereas in other cases the deposits are characterized by a high percentage of small-scale cross-lamination, lenticular laminations, and sharp-base parallel laminated sand lenses. The particular type of lithofacies is related to sediment accumulation rates, size and water depth of the interdistributary bay. The crevasse splay or bay fill deposits (Figure 4.3: Unit 9) display the typical coarsening-upward sequence associated with the infilling of the interdistributary bays. The thickness of individual units of bay fill varies considerably ranging from 3 to 10 m. Details of this particular type of deposit have been presented in the previous chapter. Often, this particular unit will be repeated numerous times, and it forms the bulk of the sand bodies within the subaerial delta deposits. Capping the entire sequence are high organic clays and peats that represent marsh deposits. These peat deposits are normally highly persistent and tend to show wide lateral continuity. Quality of the peat will vary considerably depending upon the local geochemical environment and relationship to adjacent distributaries in which fine-grained clays are often present due to river flooding.

This sequence, then, represents a typical vertical accumulation of deltaic facies in a river delta characterized by low wave energy, high percentage of fine-grained suspended sediment load, a rapidly subsiding basin, and a temperate climate. The major sand bodies within the unit are represented primarily by the distributary mouth bar deposit and the crevasse splays or bay fill deposits. Such a sequence could normally be expected to be found in ancient rock sequences in depositional environments that are characterized by these parameters. Many of the Tertiary deposits of the Gulf Coast associated with the Tertiary and Pleistocene depocenters consist of this type of deltaic facies. Hydrocarbon production from such a deltaic sequence occurs primarily in the distributary-mouth bar deposits and in the slump block sand deposits existing in the marine environment that have probably accumulated by subaqueous mass movement processes as described previously. Occasionally, small hydrocarbon accumulations have been documented from the bay fill sequences in

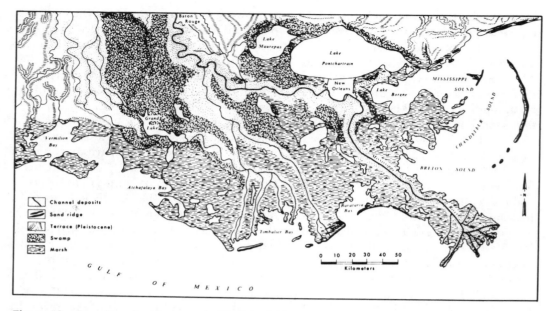

Figure 4.2. Depositional environment in Mississippi River Delta.

the upper part of the subaerial delta plain. Overall, sand deposits tend to show low lateral continuity and complex interfingering relationships. However, because of contemporaneous formation of numerous structures such as faults and slumps, reservoir-quality rocks display numerous modes of trapping mechanisms.

2. Klang River Delta

The Klang River system originates in the central highlands of Malaysia and drains into the Straits of Malacca. The drainage basin has an area of 902 km² and the river discharge averages 1100 cumecs. Sediment load consists primarily of clays and silts and only a small fraction of medium-grained sand is transported to the tide dominated Straits of Malacca. The receiving basin consists of an elongate or narrow seaway type of receiving basin, open on both ends (Figure 2.7: Type I). The most important factors affecting the Klang River Delta are marine processes, particularly because of high sedimentation rates; massive clay deposits accumulate, displaying little or no stratification. Further downcurrent the tidal flats display hummocky surfaces and consist of colonies of the brachiopod *Lingula*. These *Lingula* brachiopod flats are essentially large biomasses and density of faunal remains is extremely high. Because of the low wave action, wave reworking is generally absent and most of the faunal remains remain in growth position. Further downcurrent, *Lingula* beds are generally replaced by serpulid worms and worm rock. In these regions sedimentation rates of fine-grained clays are generally low and faunal remains make up the bulk of the deposits. The interior portions of the delta and areas lying between the meandering channels consist of broad freshwater swamps or jungles. In such tropical climates, plant growth and organic production are rapid and rate of accumulation of peat is high (0.1m/century). Peat deposits attain elevations higher than those of the bordering clastic meander belt deposits; therefore, the interdistributary areas are raised rather than depressed. Peat deposits attain considerable thicknesses ranging from 3 to 18 m. Thickness of individual peat beds generally tends to show high variability and the deposits interfinger

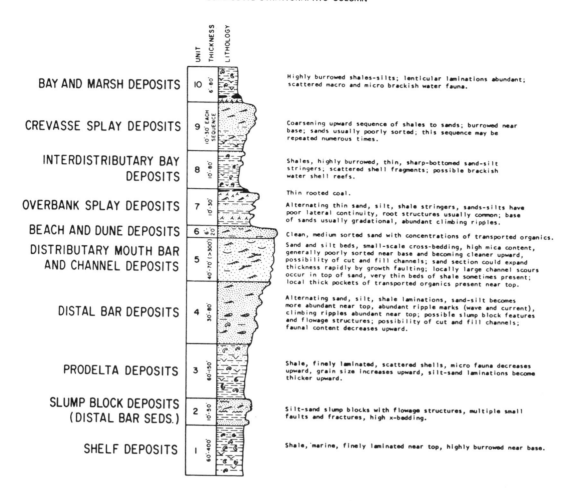

Figure 4.3. Composite stratigraphic column of Mississippi River Delta.

with tidal flat deposits and sandy deposits of the meandering tidal streams. Because of the tropical climate, woody peats tend to accumulate and peat quality is generally excellent.

The composite stratigraphic section of the Klang River Delta is illustrated in Figure 4.6. The configuration of the basin (open-ended, narrow seaway) and the presence of strong tidal currents result in tidal reworking of the sediments within the Malacca Straits which produces a well-developed tidal marine sand low in the section (Figure 4.6: Unit 2). The base of this sand is shown as sharply scoured, but this is speculative because no cores of this lower boundary were taken. However, in light of our understanding of tidal processes, there is a high probability that the basal boundary represents a scour plane. This particular sand body consists of clean well-sorted sands characterized by large-scale cross-bedding. In some instances concentrations of shell debris are found within the sand unit. Overlying the tidally-produced shelf sands are a series of clays and silts that generally form on the

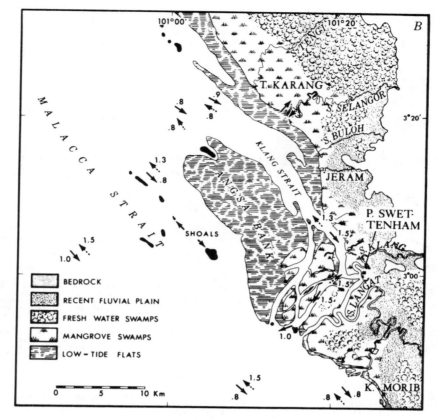

Figure 4.4. Map of Klang River Delta.

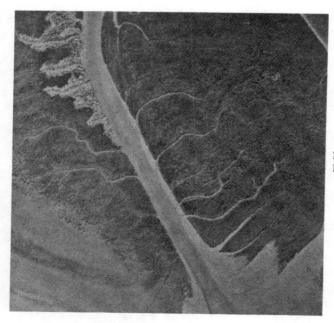

Figure 4.5. Funnel-shaped mouth of Klang River distributary.

Variability of Modern Deltas

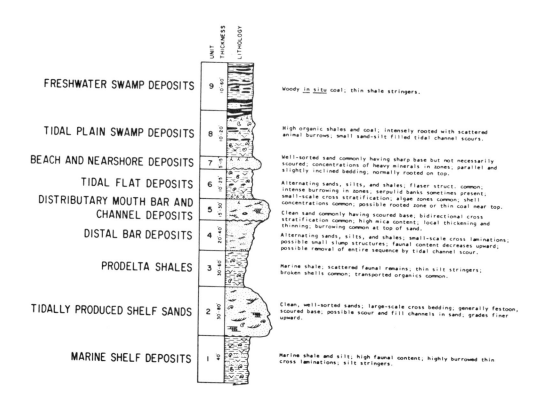

Figure 4.6. Composite stratigraphic column of Klang River Delta.

shallower portions of the delta platform, both updrift and downdrift from the major delta. These deposits (Figure 4.6: Units 3 and 4) are termed prodelta and distal bar deposits, although it is recognized that these deposits are not completely analogous to the prodelta deposits of the Mississippi River Delta. In most instances, these sediments represent fine-grained suspended material that is swept out of the river channels and lodged downdrift of the delta by tidal currents. Unit 5 (Figure 4.6) consists of distributary mouth bar deposits which, in this particular case, consist of tidally-produced sands that generally tend to show linear tidal ridge configuration, and channel deposits which represent infilling of abandoned channels by sandy deposits resulting from tidal reworking. The base of these sands displays scouring into underlying deposits. Bidirectional cross-stratification within the sand body and local thickening and thinning of units is common. In some cases, particularly in the swales separating the linear tidal ridges, concentrations of reworked shell debris can be found forming a lag concentrate. Adjacent to and interfingering with the tidal ridge deposits are the tidal flat deposits that tend to accumulate downdrift from the major delta. These deposits display high variability in lithofacies ranging from massive clays with no bioturbations to highly bioturbated silts and clays to deposits that consist primarily of faunal remains. Units 8 and 9 (Figure 4.6) represent the major subaerial delta deposits, tides, and tidal currents. Tides in the Malacca Straits are semi-diurnal with a mean spring range of 4.3 m. Tidal currents are bidirectional, but tend to set to the northwest with an average tidal current velocity of 1.5 m/sec. This constant tidal reworking of the bottom sediments in the shallow (120 to 180 m) Straits of Malacca results in the concentration of a large sand body characterized by the presence of large asymmetrical sand waves (Keller and Richards, 1967). These sand waves respond to the strong tidally produced currents, and heights of these bedforms approach 14 m. The constant reworking of the sediments by these high tidal currents results in a rather clean, well-sorted sand across much of the entire bottom of the Malacca Straits.

The major environments of deposition of the Klang River Delta are illustrated in Figure 4.4. The distributary pattern of the Klang River is extremely complex and dominated by tidal channels. The distributary mouths display broad funnel shapes and are tide dominated (Figure 4.5). A large extensive tongue-shaped bank (Angsa Bank) exists to the northwest of the main distributary mouths and is composed primarily of riverborne sediment. On this shoal and seaward of it exist large linear tidal ridges, generally oriented parallel to the axis of the strait (Off, 1963; Keller and Richards, 1967). The linear tidal ridges display high lateral dimensions and individual ridges can be traced for distances in excess of 5 to 8 km. Widths of these ridges are normally less than .5 km. These tidal ridges represent reworking and reshaping of the distributary mouth bar deposits by tidal currents into linear bodies of sand. Thus, the distributary mouth bar deposits of the Klang River Delta assume tidal characteristics. These particular deposits of the subaqueous delta have an area of 994 km^2 and are characterized by considerable bottom relief resulting from the strong bidirectional tidal currents. Within the distributaries, the strongest currents during low river are directed upstream, attaining velocities in excess of 1.5 m/sec. The channels are thus choked with well-sorted fine- to medium-grained sands in areas of maximum velocity. In other regions, not characterized by the strong tidal currents, poorly sorted clayey sands persist and tend to fill the channels. The presence of high tides and the constantly changing strong bidirectional currents results in the distributary channels of the Klang River being essentially sand filled. As a result, the distributary mouth bar (linear tidal ridges) and the channel deposits are extremely difficult to distinguish from each other and the two sand bodies generally merge. In most instances, the sands tend to display rather sharp basal contact with underlying silts and sands of the delta front deposits.

Unlike the crevasses of the Mississippi River Delta, those of the Klang River indicate that material from offshore is moved up the channel and splayed overbank by tidal processes. These overbank tidal channels leave the main river channels in an upriver-directed pattern and tend to deposit primarily fine-grained clays and silts in the adjacent tidal plains. These deposits are characterized by high organic deposition and tend to accumulate considerable thicknesses of high organic clays and peats. Adjacent to the channels are broad saline mangrove and nipa palm swamps (Watson, 1928; Coleman et al., 1970). Many of the peat deposits near the coast consist of mangrove swamps and, hence, can be characterized as a marine peat deposit. The peat deposits normally contain high quantities of sulfides. Downdrift from the mouths of the active delta and along the shoreline are broad tidal flats that display definite alongshore zonations. The tidal flats extend alongshore some 20 km northwest of the active river mouths and attain widths of 3 to 4 km. Near the mouths of the river, the tidal flats are nearly devoid of fauna and consist primarily of lower tidal plain swamp deposits. The lowermost tidal plain deposits consist primarily of highly bioturbated clays and peaty clays. The uppermost deposit consists primarily of freshwater peat deposits. In the more isolated portions of the delta plain, pure woody peats will accumulate, whereas nearer the distributaries often thin, interfingering clay stringers are common.

With both the change in basin shape and the presence of strong tidal processes, two major sands occur in the vertical sequence, both of which display marine characteristics. The high tidal range also causes most of the subaerial delta deposits to consist of tidal flat sediments. The tropical climate causes the sequence to be capped with in situ freshwater peat deposits. Thus, this stratigraphic sequence shows characteristics completely different from that of the Mississippi River Delta, which contains little or no tidally-produced deposition. These types of deltaic facies would most commonly develop in those basins characterized by narrow seaways in which tidal currents and tidal processes would tend to play a major role. Such conditions existed during Cretaceous time in portions of the western U.S. The narrow seaways associated with the Mesozoic deposits in the North Sea represent another example in which such a delta model might be applied to interpret the resulting sedimentary sequences. It is the writer's opinion that there has been a lack of recognition of tidally-produced sands in ancient rock sequences, especially in subsurface hydrocarbon-producing deposits. There is no reason to doubt that throughout geologic time, narrow seaways have existed in which tidal processes would have played the major role in producing sand bodies.

3. Ord River Delta

The Ord River debouches into one end of the structurally dominated Cambridge Gulf, a narrow elongate receiving basin oriented at a high angle to the general shoreline trend (Figure 4.7a). The river is located in western Australia in a region of extremely arid climate. Water depths within the Gulf average approximately 20 m with a maximum slightly over 80 m in depth. Tidal ranges are extreme with semi-diurnal tides having a mean range of 3.8 m and an average spring rate of 5.15 m. Within the delta proper, upstream amplification of the tidal wave as it propagates through the Gulf results in mean and spring tide ranges of 4.7 and 6.6 m respectively (Wright et al., 1973).

The Ord River rises in a tropical drainage basin 78,000 km² in area and experiences a hot, dry tropical monsoon climate. There is a water balance deficit within the drainage basin and delta of the Ord River system during every month except January and February. Thus high evaporation persists during a large part of the year. This climatic regime results in extreme annual variability in river discharge, ranging on the average from a low of 0.07 cumecs, to a maximum of 730.4 cumecs with an average discharge being 166 cumecs. Sediment load also displays extreme variability within any given month; the annual sediment load has been estimated to be 22×10^9 kg and consists of medium sand bedload and a high suspended clay-silt load.

The distributaries of the Ord River display a funnel-shaped configuration with widths of 9 km at the mouth and 0.09 km some 60 km inland (Figure 4.7b). In the lower courses of the river there is a tendency towards broad sinuous patterns, whereas in the more inland regions, extremely tight sinuous or meandering patterns are apparent. Within the lower distributary region and immediately seaward, the area is choked by sandy shoals. Linear elongate tidal ridges are aligned parallel to one another and parallel to the direction of the tidal flow within the channel (Figure 4.7b). These ridges are related to tidal bidirectional sediment transport patterns, high tidal amplitudes and tidal current symmetry. The ridges range in relief from 10 to 22 m and account for slightly over 5 to 10^6 m³ of total sand accumulation at the river mouths. The ridges show an average length of 2 km; however, those in the more seaward portions of the Cambridge and Joseph Bonaparte Gulfs can

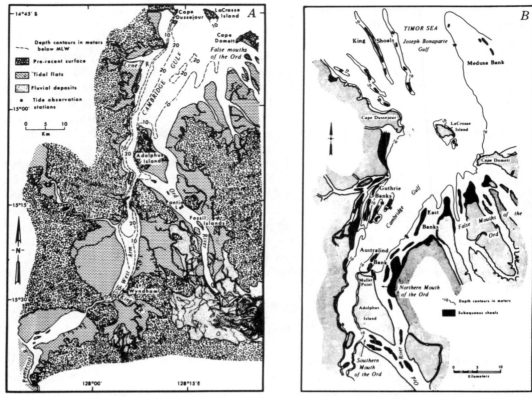

Figure 4.7. Map of Ord River Delta. A., depositional environments; B., linear tidal ridges off mouth of river.

Figure 4.8. Tidal ridge development in Ord River.

Variability of Modern Deltas

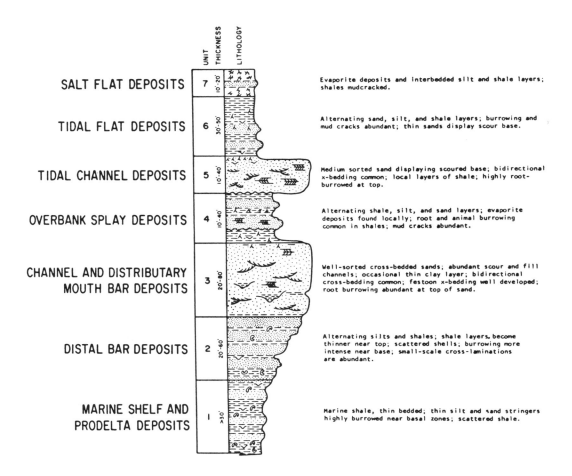

Figure 4.9. Composite stratigraphic section of Ord River Delta.

attain lengths that approach 10 to 15 km. The widths of the ridges average approximately 300 to 400 m. The bedform patterns on these ridges tend to suggest that they result from convergence of flood and ebb dominated bedload transport. Figure 4.8 is a photograph of a linear tidal ridge that occurs within the lowest portions of the distributary channels. The central part of the shoal is covered with mangrove trees which tend to stabilize its position (Figure 4.8). Along the flanks of the major ridge are seen large accumulations of sand which have been molded by migrations of large bedforms. In most instances, these linear tidal ridges form appreciable sand bodies that are composed of clean well-sorted medium-grained sands displaying external bidirectional cross-bedding.

The tidal current becomes asymmetrical within the channel itself; flood currents, being appreciably higher, attain velocities of 3.5 m/sec. In the channel proper, the larger-scale bedforms indicate upstream current directions with flood-dominated bedload transport in the shallower parts of the channel and over the shoals. Ebb flow, or seaward-directed tidal currents, tends to increase as absolute water level decreases with falling tide which causes ebb-dominated bedload transport in the deeper portions of the channels. Therefore, long periods of time are required for channel clogging by upstream transport of bedload. Through time, however, the abandoned channels tend to be filled with coarse, clean, well-sorted sands showing bidirectional cross-bedding with a strong tendency towards upstream vectors being dominant within the cross-beds, especially in the uppermost portions of the channel.

Adjacent to the major distributaries are found broad tidal flats or supratidal deposits. Fringing the distributaries, mangrove and other salt-tolerant halophytes occur and in most cases, high organic silty clays tend to form the overbank splay deposits. The interdistributary region itself attempts to build to the level of the highest tides and much of the region contains barren mud-cracked tidal flats covered by a variety of algae. The high evaporation rate results in intercalations of thin stringers of evaporites with silty-clay and sandy-clay sediments. In some instances the interior parts of the interdistributary regions often accumulate considerable quantities of evaporites. Some of the larger accumulations show dimensions of 14 by 20 km and thicknesses of up to 7 m; these accumulations having been deposited within the recent geologic past. Within the evaporite deposits, crystals and thin layers of gypsum are commonly found. Along the shoreline adjacent to the river mouths, some of the salt flats are extensive and have stranded beach ridges occurring on the tidal flats. These beach ridges are extremely thin generally and contain a high content of gypsum grains and windblown clay pellets. The low tide flats which tend to dominate this portion of the interdistributary regions are generally found at positions approximating the mean low water level. These tidal flats are composed of much sandier type material and layers of sand, silty clay and algal laminations. Branching off the main distributaries and cutting across the tidal flats are large sinuous tidal channels that transport large quantities of sandy material. These tidal channels meander actively across the delta plain and normally display quite large dimensions. Many of the channels will exceed 100 to 150 m in width and the depth of the channels will approach 10 to 15 m at high tide. The tendency to meander actively and migrate results in the formation of sandy channel deposits that range in thickness from 3 to 12 m.

The composite stratigraphic column of the Ord River Delta is illustrated in Figure 4.9. The lower most units (Figure 4.9: Units 1 and 2) represent the finer-grained deposits that are flushed out of the delta proper and accumulate in the Timor Sea, seaward of the major Joseph Bonaparte Gulf (Figure 4.7b). These deposits are generally characterized by silty and sandy clays that show high bioturbation in some cases. Small scale cross-laminations are generally common in the siltier and sandier layers. The major sand unit (Figure 4.9: Unit 3) consists of channel and distributary mouth bar deposits that are composed primarily of tidally-deposited sands. The sand deposition results from the formation of the large linear tidal ridges that were described previously; these tend to grade inland into sandy channel-filled deposits. In most instances, it is extremely difficult to separate the sandy channel fills from the tidally produced sand bodies that form the distributary mouth bar. Near the seaward edge of the sand body, large scale bidirectional cross-bedding is common. In the more inland portions of the delta, the sand shows a greater tendency towards displaying sharp scoured bases and consists primarily of landward oriented cross-bedding. The overbank splay deposits are those rather restricted deposits that form immediately adjacent to major distributaries and result from tidal overflow out of the distributaries and to the adjacent salt basins. Overall, the deposits can be described as a poorly sorted silty and sandy clay with occasional thin layers of clean well-sorted sands displaying scoured bases. These thin stringers are formed within the channels that tend to cut across the major distributary banks. In most instances high bioturbation by burrowing plants and animals is common within the deposits. The sand body (Figure 4.9: Unit 5) is composed of tidal channel sediments that consist primarily of narrow meander belt deposits; they display a shoestring sand body geometry with characteristic sharp lower basal planes indicating scour into underlying deposits. Bidirectional cross-bedding is common within the sands and herringbone types of cross-stratification are common. In most instances the top of the sand displays high bioturbation by burrowing plants. The subaerial delta deposits (Figure 4.9: Units 6 and 7) represent tidal plain sediments (supratidal flat deposits) that can be characterized by alternating sand, silt and clay layers with occasional intercalations of evaporite deposits. Occasionally the evaporite deposits tend to expand in thickness and can form large massive units.

The closed shape of the receiving basin and the feeding of the river into one end of a narrow elongate basin drastically modifies the tidal processes and results in the formation of long linear tidal sand ridges that are generally oriented perpendicular to the shoreline trend. The high tide and bidirectional tidal currents cause extensive sand infilling of the distributary channels and, as a result of the arid climate, the subaerial deposits of the delta consist primarily of supratidal and evaporite deposits. Therefore, this river delta displays sharp contrasts to the Klang and Mississippi River deltaic deposits. The lack of sand bodies showing typical riverine characteristics is noteworthy and the general lack of high organic peat-rich accumulations attests to the arid climate, which contrasts sharply with both the Mississippi River Delta and the Klang River Delta. This type of stratigraphic sequence could be expected to be found in ancient sequences in those regions characterized by internal types of drainage and along the margins of rather stable basement platforms where tidal range would be strongly amplified.

4. Burdekin River Delta

The Burdekin River drains a dry topical basin along the northeastern coast of Australia in the state of Queensland (Figure 4.10). The river debouches onto a shallow shelf behind the great barrier reef which fringes the eastern coast of Australia. The arid climate in the drainage basin results in a discharge that is extremely erratic, with a mean average discharge of 476 cumecs, but a range of 10 to 24,286 cumecs. With the onset of precipitation within the drainage basin, the river will rise suddenly within a few days, increasing the discharge from a volume of 100 cumecs to volumes greater than 4,000 cumecs. Tides along this portion of the Australian coast are semi-diurnal with an average range of 2.2 m, but average spring tides attain a range of 3.8 m. Littoral drift normally sets northward along the coast, but is not exceptionally strong and plays only a minor role in sediment transport of riverine sediment. In the deltas previously described, wave power was relatively low, generally less than 1.4×10^7 ergs/sec. Along the seaward margin of the Burdekin Delta, however, wave power is considerably higher with an average shoreline power being 6.4×10^7 ergs/sec. This one particular factor causes drastic changes in the sand body geometry orientation, and mineral composition.

The Burdekin River transports large quantities of coarse-grained material, gravels and cobbles which are abundant in the alluvial valley only a few kilometers upstream from the delta apex. The river drains a granitic region and feldspar content within the coarse clastics are common and the composition of the river sediments could be referred to as arkosic. The active river delta comprises some 430 km^2 and consists of numerous bifurcating distributaries (Figure 4.10). Within the distributaries, however, are numerous shoal areas and elongate mid-channel islands. The channels are choked by coarse sediment, a reflection of the high tidal regime. The mouths of the river display a bell-shape and sandy shoals and large bedforms are common within the lower distributaries and immediately offshore. Although

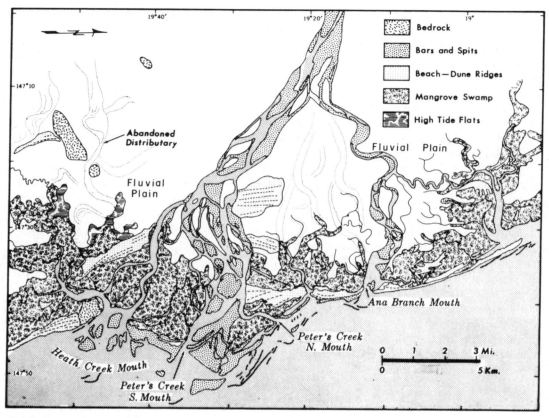

Figure 4.10. Map of Burdekin River Delta.

Figure 4.11. Aerial photograph showing beach ridges at mouth of Burdekin River distributary.

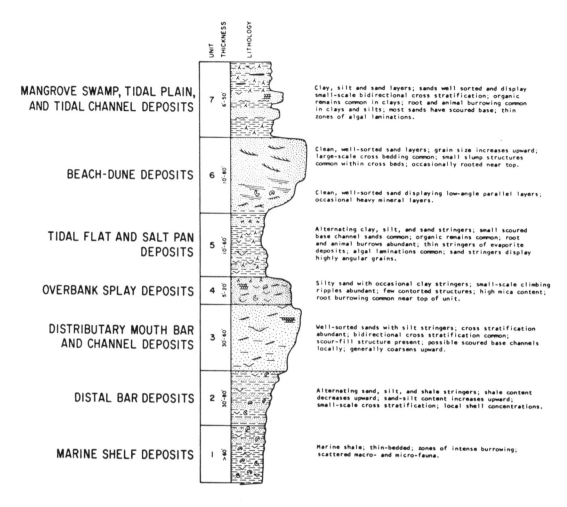

Figure 4.12. Composite stratigraphic column of Burdekin River Delta.

well-developed linear tidal ridges are not present, broad sandy shoals appear to be persistent within the nearshore waters. Thus the distributary mouth bar merges landward with sandy channel-fill deposits. Measurement of current directional properties of the sands at the river mouth tend to show bimodal distributions within the lower river course with both upstream and downstream components being persistent. At the limit of inland tidal intrusion within the channel, the dominant cross-bedding direction is strongly biased in a downstream direction. Thus, the tidal processes in this delta, as seen in the previous two examples, play an important role in the formation of distributary channel-fill sands, their directional properties, and the mechanisms responsible for infilling of the channel. Above the tidal influence, overbank crevasse splays are common and are particularly abundant at a point just above the limit of tidal inundation. These deposits consist of fine- to medium-grained sands displaying clay stringers and are generally highly bioturbated by root penetration.

The interdistributary areas consist of tidal flat deposits because of inundation by high tidal range; mangrove deposits, supratidal flats accumulations and small evaporite deposits are common within these regions. Mineral composition of all the deposits described so far shows a high percentage of feldspars and kaolinitic clays.

The major sand deposits within the river consist of beach-dune ridge deposits shown in Figure 4.10 as broad sequences of beach ridge complexes separated by narrow mangrove colonized swales. An aerial photograph (Figure 4.11) shows the relationship of these reworked sandy beach deposits in relationship to the river mouth and the swale deposits. These features result primarily from reworking and concentration of riverborne sediments by wave action. This delta is the first one described that displays appreciable quantities of marine wave-reworked sands. Within the Burdekin River, discharge is extremely erratic and major floods show a recurrence interval of approximately 10 years. During these floods, abundant sediment is delivered to the coast during a short period of time. In the intervening periods between floods, waves and tides rework and redistribute the sediments before another introduction of river sediments. This process results in alternating zones of beach ridges and swales. This particular aspect of beach ridge-swale formation is well illustrated in the vertical aerial photograph shown in Figure 4.11.

In this particular river delta, the channel sands and distributary mouth bar deposits tend to be oriented at high angles to the shoreline trend, whereas the marine reworked beach sands are oriented roughly parallel to the shoreline. Mineralogical analyses of the sands from both sand bodies tend to show appreciable differences. The channel sands show a high feldspar content, while the beach deposits approach nearly 95% quartz. Therefore, major reworking of riverine-delivered sediments to a high wave energy regime results in sorting and changing the mineral composition of the riverine sediments. The Burdekin Delta receives 6.4×10^7 ergs/sec/m of coast of wave energy annually; or in other terms, in two days more wave energy reworks this coast than along the Mississippi River Delta in 365 days. This high wave energy is responsible for reworking and cleaning the sediments and concentrating the more resistant quartz grains.

The high tide range combined with high wave energy results in the formation of broad sandy tidal flat deposits which merge locally with the beach ridge deposits themselves and in many respects the two deposits are extremely difficult to differentiate from each other. Twice daily, the shoreline transgresses and regresses an average of 3.5 km and the high wave energy reworks the tidal flat sediments into broad, well-sorted, highly quartzose, sandy tidal flat deposits. The sands are extremely wide-spread and cover an areal extent of approximately 250 km² with a sheet-like deposit of sand. In the interdistributary region, mangrove swamps, barren algal flats and localized salt pans, where evaporites are accumulating, are common deposits. Small tidal channels meander across the tidal plain and form small local concentrations of sand that rarely exceed a few meters in thickness.

The composite-stratigraphic section of the Burdekin River Delta is shown in Figure 4.12. The change in the environmental setting, particularly the increase in wave energy, results in a different distribution and genesis of sand bodies within this delta. The lowermost units represent the offshore finer-grained deposits, the marine shelf and distal bar deposits that accumulate seaward of the delta proper behind the Great Barrier Reef. The major, lowermost sand unit (Figure 4.12: Unit 3) comprises distributary mouth bar deposits and channel fill deposits. The sands are well-bedded and display a wide variety of cross-stratification with bidirectional cross-bedding being common. Scour-and-fill structures are particularly abundant in the channel deposits because of highly variable tidal currents. This particular sand body is oriented at high angles to the shoreline trend and because of the lack of wave reworking, tends to reflect the mineral composition of the sediment introduced from the drainage basin. High percentages of feldspar are common. The uppermost sand unit (Figure 4.12: Unit 6) is comprised of clean, well-sorted, high quartzose sand that forms as a result of wave reworking of deltaic sands or beach-dune deposits, that grade downward into sandy tidal flat deposits. Because of the high tidal range and the reworking of the sediments into broad tidal flats, that particular sand unit generally shows high lateral continuity. With the introduction of high wave action in a delta, the subaerial deposits begin to assume importance in terms of sand body distribution and accumulation. This is the first delta in which subaerial delta deposits consist of appreciable sandy units. The uppermost unit (Figure 4.12: Unit 7) consists of tidal plain, mangrove swamp, and tidal channel deposits. In this particular delta the tidal channel deposits do not assume any appreciable importance.

This type of delta is extremely common along many portions of the modern world shorelines. They tend to form in regions not experiencing rapid subsidence, that is, on areas with rather stable continental shelves. Facing broad expanses of open water, wave reworking of the riverine-introduced sediments becomes the dominant process. Such deltaic facies would tend to develop along rather stable continental margins such as existed in upper Paleozoic and Mesozoic times along the western U.S. Many of the upper Cretaceous deposits seen along the Colorado Front Range appear to display similar characteristics to this particular type of deltaic facies.

5. Sao Francisco River Delta

The Sao Francisco River is the second largest river system in Brazil; the river drains a basin (602,310 km^2) within the humid tropics; flow is continuous all year long. Discharge averages 3,420 cumecs with maximum and minimum of 5,818 and 1,166 cumecs, respectively. The tropical river basin experiences extreme chemical and biological weathering, thus shedding large quantities of fine-grained sediments, and the river transports an extremely high suspended sediment load. Sediment concentration within the river channel of the Sao Francisco equals or exceeds that of the Mississippi River during flood.

The river debouches into the Atlantic Ocean and has formed a triangular delta of some 735 km^2 (Figure 4.13). The bulk of the delta plain consists of beach ridge-dune deposits (Figure 4.13). Relief within the delta is considerable and even with a spring tide range of approximately 2.5 m, only 75 km^2 of the delta is subject to tidal inundation. Thus the bulk of the deposits are not subjected to tidal reworking. The Sao Francisco Delta is open to the Atlantic Ocean where the coastline displays a smooth delta shoreline with only minor protrusion at the river mouth (Figure 4.13). Offshore slope is quite steep, averaging 11.2%; as a result open ocean swell generated in the southern Atlantic Ocean is not appreciably modified and the delta shoreline is reworked by extremely high wave energy. Average wave power is 30.4 ergs/sec/m of coastline, or stated in other terms, more wave energy is expended in 10 hours on the Sao Francisco shoreline than in 365 days in the Mississippi River. This appreciable and persistent shoreline wave energy tends to produce a smooth shoreline configuration. Figure 4.14 is a low oblique aerial photograph of the Sao Francisco Delta shoreline. The river mouth, seen in the foreground, is extremely constricted, and the persistent high wave energy attempts to seal off this particular river. It is also highly probable that the high wave energy plays a major role in maintaining a single distributary channel cutting across a delta plain (Figure 4.13). Extremely large eolian dunes attaining elevations in excess of 22 m also occur (Figure 4.14). Beneath the dunes and inland on the delta plain, broad sandy beach ridges, plastered one against another, form the major landforms within the entire delta plain. The sands consist of clean, well-sorted quartz (approaching 95% to 97% quartz). Within the interior delta plain, small interweaving swales are occasionally present and sandy clays, with scattered organic materials and occasional thin evaporite deposits, are concentrated. These deposits rarely exceed a few meters in thickness.

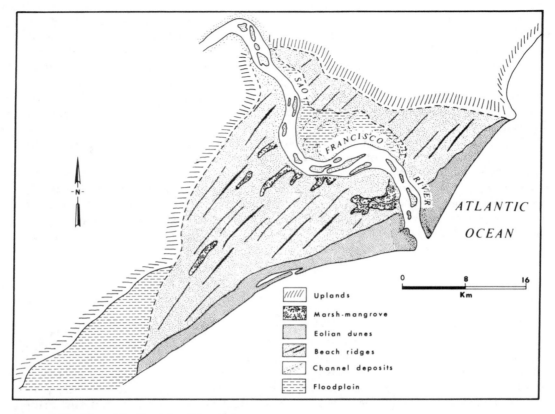

Figure 4.13. Map of Sao Francisco River Delta.

Figure 4.14. Photograph of Sao Francisco River mouth.

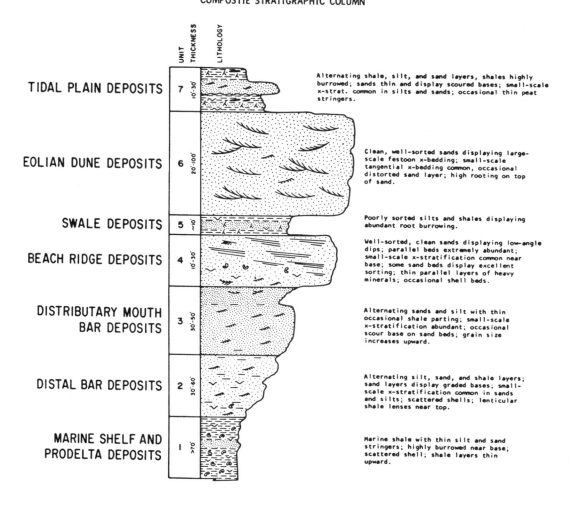

Figure 4.15. Composite stratigraphic column of Sao Francisco River Delta.

Figure 4.15 illustrates the composite stratigraphic column of the Sao Francisco Delta. The major deposits consist of beach ridge and eolian dunes. The lowermost units (Figure 4.15: Units 1, 2, and 3) form the major component part of the prograding deltaic facies. These deposits consist of sandy clays that accumulate beyond the immediate river mouths and could be termed prodelta and distal bar deposits. These grade upward into distributary mouth bar deposits which, in many cases, are extremely localized and persist only in the immediate vicinity of the river mouth. The major sand units consist of Units 4 and 6 (Figure 4.15) and are composed of beach ridge and eolian dune deposits. The beach deposits attain thicknesses of several meters to 10 m and are composed of well-sorted, clean, high-quartzose sands. The major stratification seen in exposures consists of low angle seaward-dipping accretionary, planar cross-stratification. The dune facies have been described by McKee and Bigarella (1972) from the coasts of the Sao Francisco Delta. These deposits consist of extremely thick (up to 30 m) clean, well-sorted, high-angle cross-bedded sands. The entire delta plain of the Sao Francisco River Delta consists of marine sheet sand capped by eolian dunes. As wave energy is increased in a delta region, but lacks strong littoral drift, the sand bodies tend to take on a thick sheet-like sand body geometry and are composed of clean, well-sorted quartzose sands. Thus the river-produced sand bodies that have been often ascribed to deltaic sequences are not present in this particular delta, having been replaced almost entirely by sands which result from deposits reworked by marine processes.

6. Senegal River Delta

The Senegal River drains an elongate basin on the western coast of Africa and debouches into the Atlantic Ocean near the port of St. Louis, Senegal. The river, some 1800 km long, drains a tropical basin (196,470 km² in area) with an annual rainfall between 1000 and 2050 mm. Within the delta region, however, rainfall diminishes abruptly and averages only 380 mm and evaporation is high. The annual average discharge is 770 cumecs with maximum and minimum flows of 3460 and 20 cumecs, respectively. The sediment load consists predominantly of fine-grained clays and silts with minor amounts of fine- and medium-grained sand as bedload.

The delta plain (Figure 4.16) does not protrude significantly into the receiving basin, yet it has formed a delta plain of 4250 km². The subaerial portion of the delta plain displays quite high relief and only 10% of the delta plain is inundated by the maximum spring tide range of 1.9 m. The major characteristic of this delta is its extremely high annual wave power. This delta represents one of the highest wave-dominated deltas of those studied; the average annual wave power is 112.4×10^7 ergs/sec. Stated in other terms, more wave energy reworks this delta coast in 3 hours than during 365 days in the Mississippi Delta. In addition, strong unidirectional littoral currents to the south are operative and drift velocities approaching 1 m/sec move sediment alongshore. The combination of high wave energy and strong unidirectional littoral currents results in drastically changing the distribution and orientation of sand bodies. The apex of the delta is located 160 km upstream from its mouth and flows in a generally westerly direction until it approaches the coast where it suddenly trends south-southwesterly and is deflected along the coast some 22 km by the strong southerly-flowing littoral currents. Older channels within the delta plain display a similar southward deflection (Figure 4.16).

The west, or right bank, of the river is characterized by being reworked into a large sand barrier island and the entire delta shoreline consists of large linear sand bodies. Capping the shoreline beaches are large eolian dunes, some exceeding 16 m in elevation. The entire delta plain exhibits similar geomorphic landforms: beach barrier-dune ridges oriented parallel to the coast, separated by broad organic filled river channels. This aspect of the delta plain is illustrated by the low oblique aerial photograph in Figure 4.17. Radio-carbon dating of some of the beach deposits indicates that within the past 5,000 years some 8.5 km of seaward accretion has taken place (Michel et al., 1968, 1969). Some of these barrier ridges exceed 2.5 km in width, with lengths approaching 80 km.

As mentioned previously, as the river approaches the coast, strong littoral currents deflect the river course southward and high wave action reworks the riverborne clastics into a broad active beach. With continued progradation of the river mouth parallel to the coast, the stream gradient decreases. At some point, the river will cut across its own barrier, either during floods or during hurricanes, and the process of forming a new barrier will begin. The river carries a high suspended clay load. However, the high wave energy concentrates and cleans the sand, forming beach ridges that are composed primarily of quartz-rich sand. South of the active river mouth, some of the sediments are carried downdrift and are worked onto the shoreline by the high wave action; broad beaches and large transgressive eolian sand sheets develop.

The composite stratigraphic section of the Senegal Delta is illustrated in Figure 4.18. The lowermost units (Figure 4.18: Units 1 and 2) represent the finer-grained deposits that escape from the river mouth and form only a slight subaqueous protrusion offshore from the delta proper. These types of deposits, although sandy in nature, could be referred to as prodelta and distal bar deposits. The distributary mouth bar deposit (Figure 4.18: Unit 3) is not especially well-developed and the sediments display thin shell stringers and zones of poor sorting within a sandy silt deposit. The major sands are represented by the reworked deltaic sediments, the beach-dune complex (Figure 4.18: Units 5 and 6). These coastal barriers and eolian dunes which cap the deltaic sequence are composed of clean, well-sorted sands containing an extremely high percentage of quartz. In most instances, the three sand units (Figure 4.18: Units 3, 5 and 6) are stacked one on top of another forming a single sand body in excess of 65 m thick. The sand body is normally poorly sorted and finer at the base and grades upward into clean, well-sorted sands. Unit 4 (Figure 4.18) represents the channel-fill deposits and consists of poorly-sorted organic clays. In some instances, thin peat stringers can be found within the channel fill deposits. Thus the high wave energy of this delta results in marine reworking of riverine-delivered sediments and formation of elongate marine sand bodies. Instead of being extremely widespread and sheetlike, such as described in the Sao Francisco River Delta, the sand bodies of the Senegal Delta display long narrow sandbody trends parallel to the depositional strike which are separated by clay fill channels. This orientation of the sand bodies results from the strong littoral currents combined with the high wave action. This type of sand body distribution contrasts sharply with the linear sand bodies of the Mississippi River Delta which tend to trend at high angles to the depositional strikes, protrude out into the receiving basin, and are produced solely by riverine processes. Therefore, the two extremes of deltaic sand-body types are represented by the Mississippi, a river dominated system, and the Senegal, a wave-current dominated system.

7. Summary

The various coastal processes described above exert significant control on the geometry, genesis, and distribution of deltaic facies. Although the process settings in which modern deltas are accumulating are highly variable on a worldwide basis, delta comparisons indicated that basically six types of sand distribution patterns exist in the 34 deltas studied. Figure 4.19 illustrates these six types of net sand distribution patterns. In all cases, the arrows indicate general depositional strike. Type I consists of a widespread body of sand composed primarily of distributary-mouth bar deposits. Within the sand body are definite finger-like thickenings of sands associated with individual distributaries. Thickness of the sands will depend upon the sediment load, subsidence rate, and contemporaneous deformational processes. The elongate depositional pods of sand are normally oriented at a high angle to the shoreline or depositional strike. Such a sand distribution is favored by extremely low wave energy, low tide range, low offshore slope, low littoral drift and a high fine-grained suspended sediment load. Modern deltas which tend to display this type of sand body distribution pattern include the Mississippi, the Parana, the Dneiper, and the Orinoco Deltas.

Type II sand bodies consist of finger-like protrusions of channel sands with numerous isolated sand bodies lying seaward of the shoreline. The sand fingers represent sand-filled channels and nearly always display a scored base. The linear sand bodies offshore represent deposition by tidal action and a reworking of the riverine sediments into linear tidal sands at the distributary mouth bar. In some instances, these linear sand bodies parallel generally the depositional strike (especially in narrow, open-ended seaways) and in other cases tend to trend at high angles to the general shoreline trend (in narrow seaways, closed at one end).

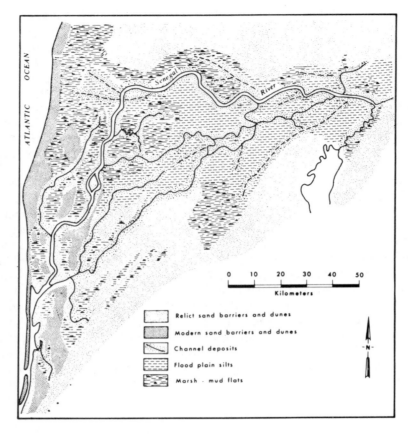

Figure 4.16. Map of Senegal River Delta.

Figure 4.17. Photograph of large beach-barriers in delta plain of Senegal River Delta.

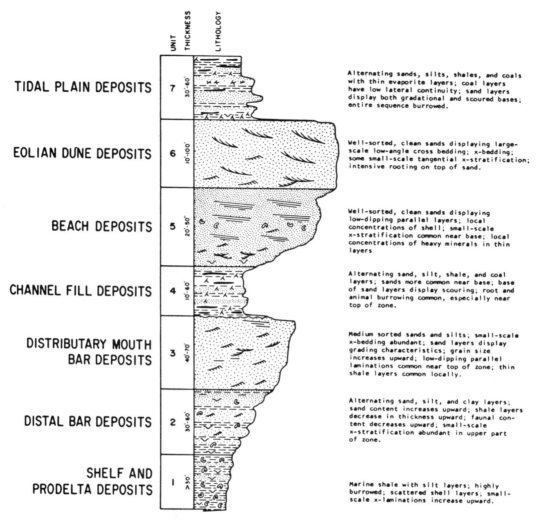

Figure 4.18. Composite stratigraphic column of Senegal River Delta.

Environmental settings conducive to the formation of this sand body pattern include low wave energy, high tide range, and narrow, restricted depositional basins which are normally indented from the coast. Modern examples of sand body distributions include the Klang, Ord, Victoria, Shatt-al-Arab, Indus, Colorado, and Ganges-Brahmaputra River Deltas.

As wave energy is increased, sand bodies begin to align with the depositional strike and yield the sand body pattern demonstrated in Figure 4.19 (Type III). Sand-filled channels still persist and form sand bodies trending at high angles to the shoreline, but the increased wave energy is responsible for reworking some of the riverine sediments and formation of beach-dune ridge complexes that parallel the shoreline and intersect the channel sands at high angles. In most cases, the sand bodies trending parallel to the depositional strike

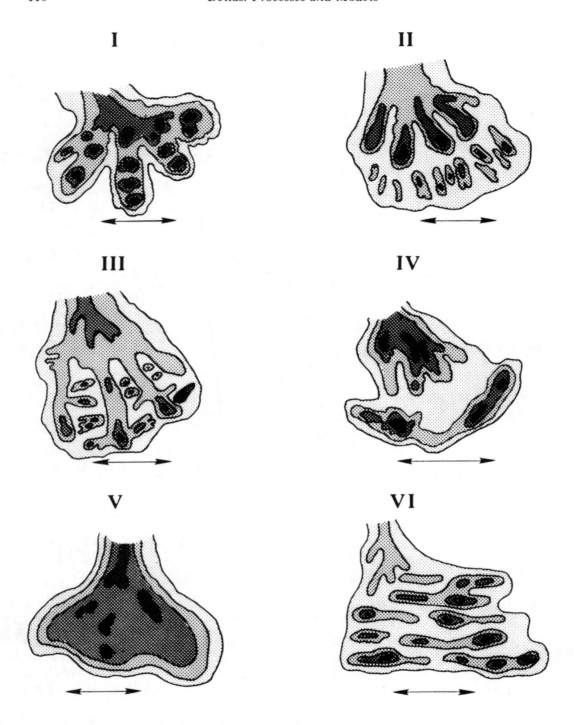

Figure 4.19. Net sand distribution patterns in modern deltas. (Republished by permission of the Houston Geological Society; Coleman and Wright, 1975.)

display higher quartz values than the adjacent channel sands. Intermediate wave energy, high tides, low littoral drift and a shallow stable receiving basin characterize this sand distribution pattern. In some instances, the channel sands will be poorly developed; this occurs as tidal range decreases. Modern day deltas illustrating this pattern include the Burdekin, Irrawaddy, Mekong, Red, and Danube River Deltas.

Several modern deltas display the pattern illustrated as Type IV in Figure 4.19; however, little is known about the processes responsible for the formation of this type. Distributary mouth bar deposits form the major finger-like sand bodies in the more landward settings. These sands, in some instances, display long finger-like protrusions; and in other instances, the distributary mouth bar deposits merge and form a widespread sheet sand with only localized thickenings. Seaward of the delta proper, beach barrier deposits have accumulated and trend parallel to the shoreline. In most instances, lagoonal deposits are located behind the offshore barrier islands. Intermediate wave energy, extremely low offshore slope and a low sediment yield tend to be associated with this pattern. Some of the modern day examples include the Appalachicola, Sagavanirktok, Horton, Rio Soto, and Brazos Deltas.

High lateral continuity of sands and development of a sheet-like geometry characterize the fifth type of sand pattern (Figure 4.19). Clean, well-sorted sands that display rather uniform thicknesses over large areas are common and localized thickening within the sand sheet is due primarily to scoured channels which may or may not be sand filled. High persistent wave energy, low littoral drift, and a steep offshore slope characterize this environmental setting. In many instances the entire sand body consists of marine reworked sands and the development of numerous beach deposits often capped by large eolian dunes. The Sao Francisco, Grijalva, Godavari, and Tana River Deltas are modern examples.

The Type VI sand pattern (Figure 4.19) is characterized by multiple large, elongate sand bodies that trend parallel to depositional strike and are separated from one another by abandoned silty clay-filled channels. The sands are often poorly sorted and dirty at the base and become more quartzose near the top. Strong littoral drift, high wave energy and steep offshore slope are associated in the formation of this pattern. The Senegal, Marowijne, and Tumpat River Deltas are modern day examples.

These patterns will likely be modified as more subsurface data are obtained, but the types presented in Figure 4.19 do exist among modern deltas. The variability within each type, however, is not known. Thus, different combinations of process setting in which particular types of deltaic sand bodies form and this sand body distribution, when compared with the vertical sequences presented in the previous sections, tends to characterize most of the modern day world deltas. It is believed they can serve as deltaic models for interpreting ancient sedimentary sequences.

REFERENCES CITED

Adams, R. D., 1970, Effects of Hurricanes Camille and Laurie on the Barataria Bay estuary: Coastal Studies Bull. 4, pp. 47-52, Louisiana State Univ., Baton Rouge.

Allen, J. R. L., 1963, Henry Clifton Sorby and the sedimentary structures of sands and sandstones in relation to flow conditions: Geol. en Mijnb., 42:223-228.

―――, 1964, Sedimentation in the modern delta of the River Niger, West Africa: In L. M. J. U. Van Straaten, ed., Deltaic and shallow marine deposits: Amsterdam, Elsevier, pp. 26-34.

―――, 1965, Fining-upward cycles in alluvial succession: Liverpool, Manchester, Geol. J., 4:229-246.

―――, 1965, Coastal geomorphology of eastern Nigeria: beach ridge, barrier islands, and vegetated tidal flats: Geol. en Mijnb., 44:1-21.

―――, 1965, Late Quaternary Niger Delta and adjacent areas: Am. Assoc. Petrol. Geologists Bull., 49:547-600.

―――, 1966, On bedforms and paleocurrents: Sedimentology, 6:153-190.

―――, 1967, Depth indicators of clastic sequences: In E. A. Hallam, ed., Depth indicators in marine sedimentary environments: Marine Geol., Special Issue 5:429-446.

―――, 1968, Current ripples—their relation to patterns of water and sediment motion: Amsterdam, North Holland Publ. Co., 433 pp.

―――, 1970, Physical processes of sedimentation—an introduction: London, G. Allen and Unwin, 248 pp.

―――, 1970, Studies in fluviatile sedimentation: a comparison of fining-upward cyclothems, with special reference to coarse-member composition and interpretation: J. Sediment. Petrol., 40:298-323.

―――, 1970, Sediments of the modern Niger Delta: In J. P. Morgan, ed., Deltaic sedimentation: modern and ancient: Soc. Econ. Paleontol. and Mineral. Spec. Publ. 15, pp. 138-151.

Andel, Tj. H. van, 1955, Sediments of the Rhone Delta, II, Sources and deposition of heavy minerals: Koninkl, Nederlandsch Geol. Mijnb. Genoot., Verh., Geol. Ser., 15:515-556.

―――, 1967, The Orinoco Delta: J. Sediment. Petrol., 37(2):297-310.

Anonymous, 1923, Changes in the Nile Delta and lower Nile; the delta lake fisheries: Geogr. Review, 13:618-620.

Arnborg, L. E., 1948, The delta of the Angerman River: Geogr. Ann., Stockholm, 30:673-690.

―――, 1959, The Lower Angerman River, Part 2, Studies of fluvial morphology and processes in the Recent delta from 1947-58 (trans.): Geogr. Inst., Uppsala Univ. (Sweden), Publ. 2, 180 pp. (in Swedish).

―――, H. J. Walker, and J. Peippo, 1962, Suspended load in the Colville River, Alaska: Coastal Studies Inst., Louisiana State Univ., Baton Rouge, Rept. 54, pp. 131-144.

*Arndorfer, D. J., 1973, Discharge patterns in two crevasses of the Mississippi River Delta: Marine Geol., 15:269-287.

Attia, M. I., 1954, Deposits in the Nile Valley and the delta: Geol. Surv. of Egypt, Cairo (Government Press).

Axelson, V., 1967, The Laitaure Delta, a study of deltaic morphology and processes: Geogr. Ann., Stockholm, 49:1-127.

Baganz, B. P., J. C. Horne, and J. C. Ferm, 1975, Carboniferous and Recent Mississippi lower delta plains: Gulf Coast Assoc. Geol. Soc. Trans., 25:183-191.

Bagnold, R. A., 1941, The physics of blown sand and desert dunes: New York, William Morrow, 265 pp.

―――, 1960, Some aspects of river meanders: U.S. Geol. Surv. Prof. Paper 283-E, pp. 135-144.

Banu, A. C., 1968, Deltas of the world (trans.): Bucharest, Natura, 20:17-26 (in Rumanian).

Barrell, J., 1906, Relative geologic importance of continental, littoral, and marine sedimentation: J. Geol., 14:316-362, 430-447, 524-568.

———, 1912, Criteria for the recognition of ancient delta deposits: Geol. Soc. Am. Bull., 23:377-466.

Barton, D. C., 1930, Deltaic coastal plain of southeast Texas: Geol. Soc. Am. Bull., 41:359-382.

*Bates, C. C., 1953, Rational theory of delta formation: Am. Assoc. Petrol. Geologists Bull., 37:2119-2162.

*Bea, R. G., 1971, How sea floor slides affect offshore structures: Oil and Gas J., 69(48):88-92.

*———, and P. Arnold, 1973, Movements and forces developed by wave induced slides in soft clays: Preprints, Offshore Tech. Conf., Houston, Texas, April 1973.

Belt, E. S., 1975, Scottish carboniferous cyclothem patterns and their paleoenvironmental significance: In M. L. Broussard, ed., Deltas, models for exploration: Houston, Texas; Houston Geol. Soc. 2nd ed. pp. 427-449.

Bernard, H. A., 1965, A resume of river delta types (abs.): Am. Assoc. Petrol. Geologists Bull., 49:334-335.

———, and C. F. Major, Jr., 1963, Recent meander belt deposits of the Brazos River: an alluvial "sand" model: Am. Assoc. Petrol. Geologists Bull., 47:350.

———, C. F. Major, Jr., B. S. Parrott, and R. J. LeBlanc, 1970, Recent sediments of southeast Texas—a field guide to the Brazos alluvial and deltaic plains and the Galveston Island complex: Texas Bur. Econ. Geol. Guidebook 11, Univ. Texas, Austin, 67 pp.

Bird, E. C. F., 1962, The river deltas of the Grippsland lakes, Part 1: Roy. Soc. Victoria Proc., 75:65-74.

Blatt, H., G. V. Middleton, and R. C. Murray, 1972, Origin of sedimentary rocks: Englewood Cliffs, New Jersey, Prentice Hall, 634 pp.

Bondar, C., 1963, Data concerning marine water penetration into the mouth of the Sulina channel: Studii de Hidraulica, 9:293-335 (in Rumanian).

———, 1967, Contact of fluvial and sea waters at the Danube River mouths in the Black Sea: Studii de Hidrologie, 19:153-164 (in Rumanian).

———, 1968, Hydraulical and hydrological conditions of the Black Sea waters penetration into the Danube mouths: Studii de Hidrologie, 25:103-120 (in Rumanian).

———, 1970, Considerations theoriques sur la dispersion d'un courant liquide de densite reduite et a niveau libre, dans un bassin contenant un liquide d'une plus grande densite: Proc. Symp. on Hydrology of Deltas, UNESCO, 11:246-256.

Bonham-Carter, G. F., and A. J. Sutherland, 1968, Diffusion and settling of sediments at river mouths: a computer simulation mouth: Gulf Coast Assoc. Geol. Soc. Trans., 17:326-338.

*Borichansky, L. S., and V. N. Mikahilov, 1966, Interaction of river and sea water in the absence of tides: In Scientific problems of the humic tropical zone deltas and their implications. UNESCO, Proc. Dacca Symp., pp. 175-180.

Bowden, K. F., 1967, Circulation and diffusion: In G. H. Lauff, ed., Estuaries: Am. Assoc. Adv. Sci. Publ. 83, pp. 15-36.

Bretschneider, C. L., 1954, Field investigation of wave energy loss in shallow water ocean waves: U.S. Army Corps of Engr. Beach Erosion Board Tech. Memo., 46:1-21.

———, 1966, Wave generation by wind, deep and shallow water: In A. T. Ippen, ed., Estuary and coastline hydrodynamics: New York, McGraw-Hill, pp. 133-196.

———, and R. O. Reid, 1954, Modification of wave height due to bottom friction, percolation and refraction: U.S. Army Corps of Engr. Beach Erosion Board Tech. Memo., 45, 36 pp.

Briscoe, C., Jr., B. Hayden, and R. Dolan, 1973, Classification of the coastal environments of the world. The climatic regimes of western South America: a case study: Dept. of Environmental Sci., Univ. of Virginia, Tech. Rept. 6, 33 pp.

Bronfman, A. M., and A. N. Aleksandov, 1965, The Don Estuary as an example of sedimentation in marine shallows in front of an estuary: Oceanology, 5:68-75 (English trans.).

Brooks, J. M., and W. M. Sackett, 1973, Sources, sinks, and concentrations of light hydrocarbons in the Gulf of Mexico: J. Geophys. Res., 78:52-58.

Broussard, M. L., ed., 1975, Deltas, 2nd ed: Houston, Texas, Houston Geol. Soc., 555 pp.

Busch, D. A., 1953, The significance of deltas in subsurface exploration: Tulsa Geol. Dig., 21:71-80.

———, 1959, Prospecting for stratigraphic traps: Am. Assoc. Petrol. Geologists Bull., 43:2829-2843.

Butzer, K. W., 1970, Contemporary depositional environments of the Omo Delta: Nature, 226:425-430.

Chien, N., 1961, The braided stream of the lower Yellow River: Sci. Sinica, Peking, 10:734-754.

———, 1967, Hydraulics of rivers and deltas: Unpub. Ph.D. dissertation, Colorado State Univ., Fort Collins, 176 pp.

Church, M. A., 1972, Baffin Island sandurs: a study of arctic fluvial processes: Geol. Surv. Canada Bull. 216, 208 pp.

Clapp, F. G., 1922, The Hwang Ho, Yellow River: Geogr. Review, 12:1-18.

Cobb, W. C., 1952, The passes of the Mississippi River: Am. Soc. Civil Engr., Preprint 13, 23 pp.

Coleman, J. M., 1966, Recent coastal sedimentation—central Louisiana coast: Coastal Studies Inst., Louisiana State Univ., Baton Rouge, Coastal Studies Series No. 17, 73 pp.

———, 1967, Deltaic evolution: *In* R. Fairbridge, *ed.*, Encyclopedia of earth sciences. New York, Reinholt, pp. 255-261.

———, 1969, Brahmaputra River: channel processes and sedimentation: Sedimentary Geol., 3:131-239.

———, 1975, Deltaic processes: *In* Finding and exploring ancient deltas in the subsurface, pp. A-1—A-25: Am. Assoc. Petrol. Geologists Marine Geol. Committee Workshop, April 6, 1975, Dallas, Texas.

*———, and S. M. Gagliano, 1964, Cyclic sedimentation in the Mississippi River deltaic plain: Gulf Coast Assoc. Geol. Soc. Trans., 14:67-80.

———, and S. M. Gagliano, 1965, Sedimentary structures—Mississippi delta plain: *In* G. V. Middleton, *ed.*, Primary sedimentary structures and their hydrodynamic interpretation—a symposium: Soc. Econ. Paleontol. and Mineral. Spec. Publ. 12, pp. 133-148.

———, S. M. Gagliano, and J. P. Morgan, 1969, Mississippi River subdeltas. Natural models of deltaic sedimentation: Coastal Studies Inst., Louisiana State Univ., Baton Rouge, Coastal Studies Bull. 3, pp. 23-27.

———, S. M. Gagliano, and W. G. Smith, 1966, Chemical and physical weathering on saline high tidal flats, northern Queensland, Australia: Geol. Soc. Am. Bull., 77:205-206.

———, S. M. Gagliano, and W. G. Smith, 1970, Sedimentation in a Malaysian high tide tropical delta: *In* J. P. Morgan, *ed.*, Deltaic sedimentation: modern and ancient: Soc. Econ. Paleontol. and Mineral. Spec. Publ. 15, 312 pp.

*———, S. M. Gagliano, and J. E. Webb, 1964, Minor sedimentary structures in a prograding distributary: Marine Geol., 1:240-258.

———, and C. L. Ho, 1968, Early diagenesis and compaction in clays: Proc. Symp. on Abnormal Subsurface Pressures, Louisiana State Univ., Baton Rouge, Nov. 1968, pp. 23-50.

———, and W. G. McIntire, 1971, Transiting coastal river channels: International Hydrographic Review, 48:11-43.

*———, D. Prior, L. E. Garrison, 1980, Subaqueous sediment instability in offshore Mississippi River Delta: (34 maps) Bureau of Land Management Open File Report 80-01, New Orleans, 63 pp.

———, and W. G. Smith, 1964, Late Recent rise of sea level: Geol. Soc. Am. Bull., 79(9): 833-840.

*———, J. N. Suhayda, T. Whelan, III, and L. D. Wright, 1974, Mass movements of Mississippi River Delta sediments: Gulf Coast Assoc. Geol. Soc. Proc., 24:49-68.

*———, and L. D. Wright, 1971, Analysis of major river systems and their deltas: procedures and rationale, with two examples: Coastal Studies Inst., Louisiana State Univ., Baton Rouge, Coastal Studies Series No. 28, 125 pp.

———, and L. D. Wright, 1973, Variability of modern river deltas: Gulf Coast Assoc. Geol. Soc. Trans., pp. 33-36.

———, and L. D. Wright, 1974, Formative mechanisms in a modern depocenter: *In* Stratigraphy and petroleum potential of northern Gulf of Mexico (part II). New Orleans Geol. Soc. Seminar, Jan. 22-24, 1974, pp. 90-139.

*———, and L. D. Wright, 1975, Modern river deltas: variability of processes and sand bodies: *In* M. L. Broussard, *ed.*, Deltas, models for exploration, 2nd ed: Houston, Texas; Houston Geol. Soc., pp. 99-150.

Collinson, J. D., 1970, Bedforms in the Tana River, Norway: Geogr. Ann., 52:31-56.

Conybeare, C. E. B., K. A. W. Crook, 1968, Manual of sedimentary structures: Bureau Mineral Resources, Geology and Geophysics, Canberra, A. C. T., Bull. No. 102, 327 pp.

Cornish, V., 1914, Waves of sand and snow: London, T. F. Univin, 383 pp.

Cotter, E., 1975, Deltaic deposits in the Upper Cretaceous Ferron Sandstone, Utah: *In* M. L. Broussard, *ed.*, Deltas, models for exploration, 2nd ed.: Houston, Texas; Houston Geol. Soc., pp. 471-484.

Credner, G. R., 1878, Die Deltas, ihre Morphologie, geographische Verbreitung, und entstehungs Bedingungen (Deltas, their morphology, geographical distribution, and growth): Petermann's Geographischen Mittheilungen, Erganzungsband 12, Erganzungsheft 56, 74 pp.

Crickmay, C. H., and C. C. Bates, 1955, Discussion of delta formation: Am. Assoc. Petrol. Geologists Bull., 39:107-114.

Crosby, E. J., 1972, Classification of sedimentary environments: *In* J. K. Rigby and W. K. Hamblin, *eds.*, Recognition of ancient sedimentary environments: Soc. Econ. and Paleontol. Mineral. Spec. Publ. No. 16, pp. 4-11.

Curray, J. R., 1960, Sediments and history of Holocene transgression, continental shelf, northwest Gulf of Mexico: Am. Assoc. Petrol. Geologists Spec. Publ., pp. 221-266.

———, 1969, Estuaries, lagoons, tidal flats and deltas: *In* D. J. Stanley, *ed.*, The new concepts of continental margin sedimentation: Am. Geol. Inst., Washington, D.C., JC-3, pp. 1-30.

———, and D. G. Moore, 1964, Pleistocene deltaic progradation of continental terrace, Costa de Nayarit, Mexico: *In* Tj. H. van Andel and G. G. Shor, Jr., *eds.*, Marine geology of the Gulf of California: Am. Assoc. Petrol. Geologists Mem. 4., pp. 193-215.

Dapples, E. C., and M. E. Hopkins, 1969, Environments of coal deposition: Geol. Soc. Am. Spec. Paper No. 114, 204 p.

Davies, D. K., 1966, Sedimentary structures and subfacies of a Mississippi River point bar. J. Geol., 74:234-239.

Davies, J. L., 1958, Wave refraction and the evolution of shoreline curves: Geographical Studies, 5(2):51-52.

Davis, W. M., 1909, The geographical cycle: In D. W. Johnson ed., Geographical essays: Dover Edition, 1954, 777 pp.

Defant, A., 1961, Physical oceanography, v. 2: New York, Macmillan, 598 pp.

Dobby, E. H. G., 1936, The Ebro Delta: Geogr. J., London, 87:455-469.

Dobson, R. S., 1967, Some applications of a digital computer to hydraulic engineering problems: Dept. Civil Engr., Stanford Univ., Tech. Rept. 80, 172 pp.

Doeglas, D. J., 1962, The structure of sedimentary deposits of braided rivers: Sedimentology, 1:167-190.

Donaldson, A. C., 1967, Deltaic sands and sandstones: In B. E. Weichman, Symp. on Recently Developed Geologic Principles and Sedimentation of the Permo-Pennsylvania of the Rocky Mountains: Wyoming Geol. Assoc., 20th Ann. Field Conf., Casper, Wyoming, 1966, Guidebook, pp. 31-62h.

*Doyle, E, H., 1973, Soil-wave tank studies of marine soil instability: Preprints, Offshore Tech. Conf., Houston, Texas, April 1973, pp. 743-766.

Duboul-Razavet, C., 1956, Contribution a l'etude geologique et sedimentologique du delta du Rhône: Soc. Geol. France Mem. 76, 35, 234 pp.

Edelman, C. H., 1956, Sedimentology of the Rhine and Meuse deltas as an example of the sedimentology of the Carboniferous: Geol. en Mijnb., Geol. Ser. 16, pp. 64-75.

Egorov, V. V., 1966, Main structural features of the great deltas in the territory of the U.S.S.R.: In Scientific problems of the humid tropical zone deltas and their implications: UNESCO, Proc. Dacca Symp., pp. 393-397.

Ellison, T. H., and J. S. Turner, 1959, Turbulent entrainment in stratified flows: J. Fluid Mech., 6:423-448.

Ethridge, F. G., T. R. Gopinath, and D. K. Davies, 1975, Recognition of deltaic environments from small samples: In M. L. Broussard, ed., Deltas, models for exploration, 2nd ed.: Houston, Texas; Houston Geol. Soc., pp. 151-164.

Everett, D. K., 1971, Hydrologic and quality characteristics of the lower Mississippi River: Louisiana Dept. Public Works, 48 pp.

Fahnestock, R. K., 1963, Morphology and hydrology of a braided stream: U. S. Geol. Surv. Prof. Paper 422A, pp. 1-70.

Farmer, H. G., and G. W. Morgan, 1953, The salt wedge: Proc., 3rd Conf. on Coastal Eng., pp. 54-64.

Fisher, W. L., and L. F. Brown, 1972, Clastic depositional systems—a genetic approach to facies analysis: Bur. of Econ. Geol., Univ. of Texas at Austin, 211 pp.

———, and J. H. McGowen, 1969, Depositional systems in Wilcox Group (Eocene) of Texas and their relation to occurrence of oil and gas: Am. Assoc. Petrol. Geologists Bull., 53:30-54.

———, L. F. Brown, Jr., A. J. Scott, and J. H. McGowen, 1969, Delta systems in the exploration for oil and gas, a research colloquium: Bur. of Econ. Geol., Univ. of Texas at Austin, 78 pp.

Fisk, H. N., 1944, Geological investigation of the alluvial valley of the lower Mississippi River: U. S. Army Corps of Engr., Mississippi River Commission, Vicksburg, Miss.

———, 1947, Fine-grained alluvial deposits and their effects on Mississippi River activity: U. S. Army Corps of Engr., Mississippi River Commission, Vicksburg, Miss.

———, 1948, Geological investigations of the lower Mermentau River basin and adjacent areas in coastal Louisiana: U. S. Army Corps of Engr., Mississippi River Commission, Vicksburg, Miss.

———, 1952, Geological investigation of the Atchafalaya Basin and problems of Mississippi River diversion: U. S. Army Corps of Engr., Mississippi River Commission, Vicksburg, Miss., pp. 1-145.

———, 1955, Sand facies of Recent Mississippi Delta deposits: Proc., 4th World Petrol. Congr., Rome, Sec. 1, pp. 377-398.

———, 1956, Nearshore sediments of the continental shelf off Louisiana: Proc. 8th Texas Conf. on Soil Mech. and Foundation Engr., pp. 1-23.

———, 1960, Recent Mississippi River sedimentation: In E. Aelst, ed., 4th Congres pour l'avancement etudes de stratigraphie et de geologie du Carbonifere, Heerlen, Compte Rendu, 1:187-199.

*———, 1961, Bar-finger sands of the Mississippi Delta: In J. A. Peterson and J. C. Osmond, eds., Geometry of sandstone bodies: Am. Assoc. Petrol. Geologists, Tulsa, Okla., pp. 29-52.

———, and E. McFarlan, Jr., 1955, Late Quaternary deltaic deposits of the Mississippi River—local sedimentation and basin tectonics: In A. Poldevaart, ed., Crust of the earth, a symposium: Geol. Soc. Am. Spec. Paper 62, pp. 279-302.

———, and E. McFarlan, Jr., 1959, Recent peat deposits, Louisiana coastal plain: In Environments of coal deposition. Geol. Soc. Am. Spec. Paper 114, pp. 63-85.

———, E. McFarlan, Jr., C. R. Kolb, and L. J. Wilbert, Jr., 1954, Sedimentary framework of the modern Mississippi Delta. J. Sediment. Petrol., 24:76-99.

Fleming, R. H., 1940, The composition of plankton and units for reporting population and production: Proc., 6th Pacific Sci. Congr., San Francisco, Calif., 1939, pp. 535-540.

Folk, R. L., 1966, A review of grain-size parameters: Sedimentology, 6:73-93.

———, 1968, 1974, Petrology of sedimentary rocks: Austin, Texas; Hemphill's.

Frank, D. S., W. M. Sackett, R. Hall, and A. D. Fredericks, 1970, Methane, ethane, and propane concentrations in the Gulf of Mexico: Am. Assoc. Petrol. Geologists Bull., 54:1933.

*Frazier, D. E., 1967, Recent deltaic deposits of the Mississippi River, their development and chronology; Gulf Coast Assoc. Geol. Soc. Trans., 17:287-315.

———, and A. Osanik, 1961, Point-bar deposits, Old River Locksite, Louisiana: Gulf Coast Assoc. Geol. Soc. Trans., 11:121-137.

Frey, R. W., 1971, Ichnology—the study of fossil and recent lebensspuren: In B. F. Perkins, ed., Trace fossils, a field guide to selected localities in Pennsylvanian, Permian, Cretaceous and Tertiary of Texas and related papers: School of Geosciences, Louisiana State Univ., Baton Rouge, Misc. Publ. No. 71-1, pp. 91-125.

Friedman, G. M., 1961, Distinction between dune, beach and river sands from their textural characteristics: J. Sediment. Petrol., 31:514-529.

Gade, H. G., 1958, Effects of a non-rigid, impermeable bottom on plane surface waves in shallow water: J. Marine Res., 16:61-82.

Gagliano, S. M., and J. L. van Beek, 1970, Hydrologic and geologic studies of coastal Louisiana: In Geology and geomorphic aspects of deltaic processes, Mississippi Delta system: Coastal Studies Inst. and Dept. Marine Sci., Louisiana State Univ., Baton Rouge, v. 1, part 1, 140 pp.

Gagliano, S. M., and W. G. McIntire, 1968 Reports on the Mekong River delta: Coastal Studies Inst., Louisiana State Univ., Baton Rouge, Tech, Rept. 57, 143 pp.

Galloway, W. E., 1975, Process framework for describing the morphological and stratigraphic evolution of deltaic depositional systems: In M. L. Broussard, ed., Deltas, models for exploration, 2nd ed.: Houston, Texas: Houston Geol. Soc., pp. 87-98.

*Garrison, L. E., 1974, The instability of surface sediments on parts of the Mississippi Delta front: U. S. Geol. Surv. Open File Rept. Corpus Christi, Texas, 18 pp.

Gellert, J. F., 1968, The Yangtse River mouth and delts (trans.): Warsaw, Przeglad Geog., 40:413-418 (in German).

Gibbs, R. J., 1970, Circulation in the Amazon River estuary and adjacent Atlantic Ocean: J. Marine Res., 28:113-123.

Gilbert, G. K., 1899, Ripple marks and cross-bedding: Geol. Soc, Am. Bull., 10: 135-140.

———, 1914, The transportation of debris by running water; U. S. Geol. Surv. Prof. Papers, v. 86, 263 pp.

Glennie, K. W., 1970, Desert sedimentary environments: Developments in Sedimentology, No. 14, Amsterdam, Elsevier, 222 pp.

Gould, H. R., 1970, The Mississippi Delta complex: In J. P. Morgan, ed., Deltaic sedimentation: modern and ancient: Soc. Econ. Paleontol. and Mineral. Spec. Publ. 15, pp. 3-30.

Gorsline, D. S., 1967, Sedimentologic studies of the Colorado Delta: Univ. of Southern California, Geol. Rept. 67-1, 121 pp.

Greenman, N. N., and R. J. LeBlanc, 1956, Recent marine sediments and environments of northwest Gulf of Mexico: Am. Assoc. Petrol. Geologists Bull., 40:813-847.

Gwinn, V. E., 1964, Deduction of flow regime from bedding characteristics in conglomerates and sandstones: J. Sediment. Petrol., 34:656-658.

Hack, J. T., 1957, Studies of longitudinal stream profiles in Virginia and Maryland: U. S. Geol. Surv. Prof. Paper 294-B, pp. 45-97.

———, 1960, Interpretation of erosional topography in humid temperate regions: Am. J. Sci., 258A:80-97.

Harms, J. C., 1969, Hydraulic significance of some sand ripples: Geol. Soc. Am. Bull., 30:363-396.

———, D. B. MacKenzie, and D. G. McCubbin, 1963, Stratification in modern sands of the Red River, Louisiana: J. Geol., 71:556-580.

———, J. B. Southard, D. R. Spearing, and R. G. Walker, eds., 1975, Depositional environments as interpreted from primary sedimentary structures and stratification sequences: Lecture Notes, Soc. Econ. Paleontol. and Mineral. Short Course No. 2, Dallas, Texas, 161 pp.

Hayes, M. O., 1967, Relationship between coastal climate and bottom sediment type on the inner continental shelf: J. Marine Geol., 5:111-132.

———, 1967, Hydrographic control of sediment patterns and environments on depositional coasts: a model for paleogeographic reconstruction: Proc. 7th Internat. Sedimentol. Congr., Edinburgh.

———, ed., 1969, Coastal environments: N. E. Massachusetts and New Hampshire: Cont. No. 1-CRG, Dept. of Geol. Publ. Series, Univ. of Massachusetts, 462 pp.

Hedberg, H. D., 1974, Relation of methane generation to undercompacted shales, shale diapirs, and mud volcanoes: Am. Assoc. Petrol. Geologists Bull., 58(4):661-673.

*Henkel, D. J., 1970, The role of waves in causing submarine landslides: Geotechnique, 20:75-80.

Henry, V. R., 1961, Recent sedimentation and related oceanographic factors in the west Mississippi Delta area: Unpub. Ph.D. thesis, Texas A & M University.

Hickin, E. J., 1974, The development of meanders in natural river channels: Am. J. Sci., 274:414-442.

Ho, C. L., and J. M. Coleman, 1969, Consolidation and cementation of Recent sediments in the Atchafalaya Basin: Geol. Soc. Am. Bull., 80:185-192.

Hobday, D. K., and D. Mathew, 1975, Late Paleozoic fluviatile and deltaic deposits in the northeast Karroo Basin, South Africa: In M. L. Broussard, ed., Deltas, models for exploration, 2nd ed.: Houston, Texas; Houston Geol. Soc., pp. 457-469.

Holle, C. G., 1952, Sedimentation at the mouth of the Mississippi River: 2nd Conf. on Coastal Eng., New Orleans Dist. Corps of Engr., pp. 111-129.

Hollerwoger, F., 1965, The accelerated growth of river deltas in Java: Madjalah Geografi Indonesia, 4:1-15 (English ed.).

———, 1966, The progress of the river deltas in Java: In Scientific problems of the humid tropic zone deltas and their implications: UNESCO, Proc. Dacca Symp., pp. 347-355.

Horton, R. E., 1945, Erosional development of streams and their drainage basins: Geol. Soc. Am. Bull., 56:275-370.

Howe, H. V., and C. K. Moresi, 1931, Geology of Iberia Parish, Louisiana: Louisiana Geol. Surv., Geol. Bull. 1, 187 pp.

Hsu, S. A., 1971, Wind stress criteria in eolian sand transport: J. Geophys. Res., 76:8684-8686.

———, 1974, Computing eolian sand transport from shear velocity measurements: J. Geol., 81:739-743.

Hubert, J. F., 1972, "Shallow-water" prodelta flysch-like sequence in the upper Cretaceous deltaic rocks, Wyoming, and the problem of the origin of graded sandstones. In Proceedings, 24th International Geological Congress, Montreal, 1972, Section 6, Stratigraphy and Sedimentology: Geological Survey of Canada, Ottawa., pp. 107-114.

Humphreys, M., and G. M. Friedman, 1975, Upper Devonian catskill deltaic complex in north-central Pennsylvania: In M. L. Broussard, ed., Deltas, models for exploration, 2nd ed.: Houston, Texas; Houston Geol. Soc., pp. 369-379.

Ichiye, T., 1953, On the effects of waves on the vertical distribution of water temperatures: Records Oceanog. Works Japan. v. 1.

Imbrie, J., and H. Buchanan, 1965, Sedimentary structures in modern carbonate sands of the Bahamas: In G. V. Middleton, ed., Primary sedimentary structures and their hydrodynamic interpretation: Soc. Econ. Paleontol. and Mineral. Spec. Publ. No. 12, pp. 149-172.

Inman, D. L., 1952, Measures for describing the size distribution of sediments: J. Sediment. Petrol., 22:125-145.

*———, and C. E. Nordstrom, 1971, On the tectonic and morphologic classification of coasts: J. Geol., 79:1-21.

Ippen, A. T., and G. B. Keulegan, 1965, Salinity intrusions in estuaries: In C. F. Wicker, ed., Evaluation of present state of knowledge of factors affecting tidal hydraulics and related phenomena: U. S. Army Corps of Engr., Comm. Tidal Hydraulics, Chap. 4.

Jackson, R. G., II, 1973, Velocity-bedform-texture pattern of meander bends in the Lower Wabash River (abs.): Geol. Soc. Am. Abs. with Programs, 5:681.

———, 1975, A depositional model of point bars in the Lower Wabash River meander belt: Unpub. Ph.D. thesis, Univ. of Illinois, 269 pp.

———, 1975, Velocity-bedform-texture patterns of meander bends in the Lower Wabash River: Geol. Soc. Am. Bull., 86:1151-1522.

———, 1975, Hierarchical attributes and a unifying model of bedforms composed of cohesionless material and produced by shearing flow: Geol. Soc. Am. Bull., 86:1523-1533.

Jahns, R. H., 1947, Geologic features of the Connecticut Valley, Massachusetts, as related to recent floods: U. S. Geol. Surv. Water Supply Papers, 996:1-158.

Jennings, J. N., 1955, The influence of wave action on coastal outline in plan: Austrian Geographer, 6:36-44.

Johnson, J. W., 1960, The effect of wind and wave action on the mixing and dispersion of wastes: Proc., 1st Internat. Conf. on Waste Disposal in the Marine Environment: New York; Pergamon, pp. 328-343.

———, and P. S. Eagleson, 1966, Coastal processes: In A. T. Ippen, ed., Estuary and coastline hydrodynamics: New York; McGraw-Hill, pp. 404-492.

———, and H. C. Hwang, 1961, Mixing and dispersion by wind waves: Inst. Eng. Res., Univ. of California, Tech. Rept. Ser. 138, Issue 5.

Johnston, W. A., 1921, Sedimentation of the Fraser River Delta: Canada Geol. Surv. Mem., 125:1-46.

———, 1922, The character of the stratification of the sediments in the Recent delta of the Fraser River Delta, British Columbia, Canada: J. Geol., 30:115-129.

Jopling, A. V., 1960, An experimental study on the mechanics of bedding: Unpub. Ph.D. dissertation, Harvard Univ.

*———, 1963, Hydraulic studies on the origin of bedding: Sedimentology, v. 2.

———, 1965, Hydraulic factors and the shape of laminae: J. Sediment. Petrol., 35:777-791.

———, and E. V. Richardson, 1966, Backset bedding developed in shooting flow in laboratory experiments: J. Sediment. Petrol., 36:821-825.

Kanes, W. H., 1970, Facies and development of the Colorado River delta in Texas: In J. P. Morgan, ed., Deltaic sedimentation: modern and ancient: Soc. Econ. Paleontol. and Mineral. Spec. Publ. 15, pp. 78-106.

Kashiwamura, M., and S. Yoshida, 1967, Outflow pattern of fresh water issued from a river mouth: Coastal Eng. Japan, 10:109-115.

———, 1969, Flow pattern of density current at a river mouth: Proc., 13th Internat. Assoc. Hydraulic Res. Congr. 3:181-190.

———, 1971, Transient acceleration of surface flow at a river mouth: Coastal Eng. Japan, 14:135-142.

*Keller, G. H., and A. F. Richards, 1967, Sediments of the Malacca Straits, southeast Asia: J. Sediment. Petrol., 37:102-127.

Keulegan, G. H., 1949, Interfacial instability and mixing in stratified flows: J. Res. NBS, 43:487-500.

Klein, G. deV., 1967, Paleocurrent analysis in relation to modern sediment dispersal patterns: Am. Assoc. Petrol. Geologists Bull., 51:366-382.

———, 1975, Sandstone depositional models for explanation for fossil fuels: Champaign, Illinois Continuing Education Publ. Co., 108 pp.

———, U. DeMelo, and J. C. Della Favera, 1976, Subaqueous gravity processes on the front of Cretaceous deltas, Reconcavo Basin, Brazil: Geol. Soc. Am. Bull., 83:1469-1492.

Kolb, C. R., and W. K. Dornbusch, 1975, The Mississippi and Mekong Deltas—a comparison: *In* M. L. Broussard, *ed.*, Deltas, models for exploration, 2nd ed: Houston, Texas; Houston Geol. Soc., pp. 193-207.

———, and J. R. Van Lopik, 1958, Geology of Mississippi River deltaic plain, southeastern Louisiana: U. S. Army Corps of Engr., Waterways Experiment Station, Tech. Rept. 3-483 and 3-484, 2 volumes, Vicksburg, Miss.

———, and J. R. Van Lopik, 1966, Depositional environments of the Mississippi River deltaic plain, southeastern Louisiana: *In* M. L. Shirley and J. A. Ragsdale, *ed.*, Deltas: Houston, Texas; Houston Geol. Soc., pp. 17-62.

Komar, P. D., 1973, Computer models of delta growth due to sediment input from rivers and longshore transport: Geol. Soc. Am. Bull., 84:2217-2226.

———, and D. L. Inman, 1970, Longshore sand transport on beaches: J. Geophys. Res., 75:5914-5927.

Krigstrom, A., 1962, Geomorphological studies of sandur plains and their braided rivers in Iceland: Geogr. Ann., Stockholm, 44: 328-346.

Kruit, C., 1951, Sediments of the Rhone Delta: Granenbrage, Mouton.

———, 1955, Sediments of the Rhone Delta, 1, Grain size and microfauna: Kon. Nederlands Geol. Mijnb. Gen. Verhand., 15:397-499.

———, 1963, Is the Rhine Delta a delta?: Kon. Nederlands Geol. Mijnb. Gen. Verhand., 21:259-266.

Kuiper, E., 1960, Sediment transport and delta formation: J. Am. Soc. Civil Engr., Hydraulics Div., 86 (HY2), Part 1, pp. 55-68

Kukal, Z., 1970, Geology of Recent sediments: Prague (Academia), 490 pp.

Lagaaij, R., and F. P. H. W. Kopstein, 1964 Typical features on a fluviomarine offlap sequence. *In* L. M. J. U. van Straaten, *ed.*, Developments in Sedimentology, v. 1, Deltaic and shallow marine deposits: Amsterdam, Elsevier, pp. 216-226.

Langbein, W. B., and S. A. Schumm, 1958, Yield of sediment in relation to mean annual precipitation: Am. Geophys. Union Trans. 39:1076-1084.

Larras, J., 1957, Plages et cotes de Sable: Paris (Eyrolles), 117 pp.

LaBlanc, R. J., 1972, Geometry of sandstone reservoir bodies: *In* T. D. Cook, *ed.*, Underground waste management and environmental implications: Am. Assoc. Petrol. Geologists Mem. 18, pp. 133-190.

———, 1975, Significant studies of modern and ancient deltaic sediments: *In* M. L. Broussard, *ed.*, Deltas, models for exploration, 2nd ed: Houston Texas, Houston, Geol. Soc., pp. 13-85.

Leighly, J., 1934, Turbulence and the transport of rock debris: Geogr. Review, 24: 453-464.

Leopold, L. B., and W. B. Langbein, 1966, River meanders: Sci. Am., 214:60-70.

———, and T. Maddock, Jr., 1953, The hydraulic geometry of stream channels and some physiographic implications: U. S. Geol. Surv. Prof. Paper 252, pp. 1-57.

———, and M. G. Wolman, 1957, River channel patterns: braided, meandering, and straight: U. S. Geol. Surv. Prof. Paper 282-B, pp. 29-85.

———, and M. G. Wolman, 1960, River meanders: Geol. Soc. Am. Bull., 71: 769-794.

———, M. G. Wolman, and J. P. Miller, 1964, Fluvial processes in geomorphology: San Francisco, W. H. Freeman, 522 pp.

Lewin, J., 1976, Initiation of bedforms and meanders in coarse-grained sediment: Geol. Soc. Am. Bull., 87:281-285.

Liteanu, E., and A. Pricajan, 1963, The geological structure of the Danube Delta (trans.): Hidrobiologia, Bucharest, 4:57-82 (in Rumanian).

Lohse, E. A., 1955, Dynamic geology of the modern coastal region, northeast Gulf of Mexico: *In* J. L. Hough and H. W. Menard, *eds.*, Finding ancient shorelines, a symposium: Soc. Econ. Paleontol. and Mineral. Spec. Publ. 3, pp. 99-105.

Mackay, J. R., 1963, The Mackenzie Delta area, N.W.T.: Canada Geol. Surv., Geogr. Branch, Ottawa, Mem. 8, 202 pp.

Maldonado, A., 1975, Sedimentation, stratigraphy, and the development of the Ebro Delta, Spain: *In* M. L. Broussard, *ed.*, Deltas, models for exploration, 2nd ed.: Houston, Texas. Houston Geol. Soc., pp. 311-338.

Martens, C. S., and R. A. Berner, 1974, Dissolved gases in anoxic Long Island Sound interstitial waters (abs.): Am. Geophys. Union, 55:319.

Martin, R. H., 1967, The Recent delta of the Guadalupe River, Texas: Unpub. Ph.D. dissertation, West Virginia Univ.

Matthews, R. K., 1974, Dynamic Stratigraphy Englewood Cliffs, New Jersey, Prentice Hall, 367 pp.

Matthews, W. H., and F. P. Shepard, 1962, Sedimentation of Fraser River Delta, British Columbia: Am. Assoc. Petrol. Geologists Bull. 46:1416-1438.

McEwen, M. C., 1963, Sedimentary facies of the Trinity River Delta: Unpub. Ph.D. dissertation, Rice Univ., Houston, Texas, 113 pp.

———, 1969, Sedimentary facies of the modern Trinity Delta: In M. L. Broussard, ed., Holocene geology of the Galveston Bay area: Houston, Texas; Houston Geol. Soc., pp. 53-77.

McGowen, J. H., 1970, Gum Hollow fan delta, Nueces Bay, Texas: Bur. Econ. Geol., Univ. of Texas, Rept. of Investigation No. 69, 91 pp.

McKee, E. D., 1939, Some types of bedding in the Colorado River Delta: J. Geol., 47:64-81.

———, 1957, Primary structures in some Recent sediments. Am. Assoc. Petrol. Geologists Bull., 41:1704-1747.

*———, and J. J. Bigarella, 1972, Deformational structures in Brazillian coastal dunes: J. Sediment. Petrol., 42:670-681.

———, and G. W. Weir, 1953, Terminology for stratification and cross stratification in sedimentary rocks: Geol. Soc. Am. Bull., 64: 381-390.

Meade, R. H., 1966, Factors influencing the early states of the compaction of clays and sands—a review: J. Sediment. Petrol., 36: 1085-1101.

Meckel, L. D., 1975, Holocene sand bodies in the Colorado Delta, Salton Sea, Imperial County, California: In M. L. Broussard, ed., Deltas, models for exploration, 2nd ed: Houston, Texas; Houston Geol. Soc., pp. 239-265.

Melton, M. A., 1957, An analysis of the relations among elements of climate, surface properties, and geomorphology: Dept. Geol., Columbia Univ., New York, Tech Rept. 11, 102 pp.

Metcalf, W. G., 1968, Shallow currents along the northeastern coast of South America: J. Marine Res., 26:232-243.

*Michel, P., P. Elounard, and H. Faure, 1968, Nouvelles recherches sur le Quaternaire recent de la region de Saint-Louis (Senegal) (New research on the Recent Quaternary of the Saint Louis region, Senegal): Bull. Inst. Fundamental d'Afrique Noire, Ser. A., 30: 1-38.

*Michel, P., P. Elounard, and H. Faure, 1969, Nouvelle chronolgie des depots du Quaternaire de la region de Saint-Louis (Senegal) (New chronology of the Quaternary deposits of the Saint Louis region, Senegal): Bull. Inst. Fundamental d'Afrique Noire, Ser. A., 31-231-234.

Middleton, G. V., ed., 1965, Primary sedimentary structures and their hydrodynamic interpretation: Soc. Econ. Paleontol. and Mineral. Spec. Publ. 12, 265 pp.

———, 1965, Antidune crossbedding in a large flume: J. Sediment. Petrol., 35:922-927.

Mitchell, R. J., K. K. Tsui, and D. A. Sangrey, 1972, Failure of submarine slopes under wave action: Proc., 13th Coastal Eng. Conf., Vancouver, B.C., Canada, pp. 1515-1541.

Mikahilov, V. N., 1966, Hydrology and formation of river mouth bars: In Scientific problems of the humid tropic zone deltas and their implications. UNESCO, Proc. Dacca Symp. pp. 59-64.

———, 1971, Dynamics of the flow and the bed in nontidal river mouths: Moscow, 259 pp. (in Russian).

Moore, B. J., and R. D. Shrewsbury, 1966, Analysis of natural gases of the United States: U.S. Bur. of Mines, Pittsburgh, Pa., Information Circular 8302.

Moore, D. G., 1966, Deltaic sedimentation: Earth Sci. Reviews, 1:87-104.

———, and P. G. Scruton, 1957, Minor internal structures of some recent unconsolidated sediments: Am. Assoc. Petrol. Geologists Bull., 41:2723-2751.

Moore, G. D., 1961, Submarine slides. J. Sediment. Petrol., 31:343-357.

Moore, G. T., 1970, Role of salt wedge in bar-finger sand and delta development: Am. Assoc. Petrol. Geologists Bull., 54:326-333.

———, and D. O. Asquith, 1971, Delta: term and concept. Geol. Soc. Am. Bull., 82:2563-2568.

Moore, T. C., Jr., Tj. H. van Andel, W. H. Blow, and G. R. Heath, 1970, Large submarine slide off northeastern continental margin of Brazil: Am. Assoc. Petrol. Geologists Bull., 54:125-128.

Morgan, J. P., 1951, Report on the mudlumps at the mouths of the Mississippi River, Part 1, The occurrence and origin of the mudlumps at the mouths of the Mississippi River: New Orleans District, U. S. Army Corps of Engr., Unpub. Rept. 127 pp.

*———, 1961, Mudlumps at the mouths of the Mississippi River: In Genesis and paleontology of the Mississippi River mudlumps: Louisiana Dept. of Conservation, Geol. Bull. 35, pp. 1-116.

———, 1967, Ephemeral estuaries of the deltaic environment: In Lauff, G. H. ed., Estuaries, Am. Assoc. Adv. Sci. monograph, pp. 115-120.

———, ed., 1970, Deltaic sedimentation, modern and ancient: Soc. Econ. Paleontol. and Mineral. Spec. Publ. 15, 312 pp.

*———, J. M. Coleman, and S. M. Gagliano, 1963, Mudlumps at the mouth of South Pass, Mississippi River: sedimentology, paleontology, structure, origin, and relation to deltaic processes: Coastal Studies Inst., Louisiana State Univ., Baton Rouge, Coastal Studies Series No. 10, 190 pp.

*———, J. M. Coleman, and S. M. Gagliano, 1968, Mudlumps: diapiric structures in Mississippi Delta sediments: In J. Braunstein and G. D. O'Brien, eds., Diapirism and diapirs, pp. 145-161, Am. Assoc. Petrol. Geologists Mem. 8.

———, and P. B. Larimore, 1957, Changes in the Louisiana shoreline: Gulf Coast Assoc. Geol. Soc. Trans., 17:303-310.

———, L. G. Nichols, and M. Wright, 1958, Morphological effects of Hurricane Audrey on the Louisiana coast: Coastal Studies Inst., Louisiana State Univ., Baton Rouge, Tech. Rept. 10A, 53 pp.

———, and R. H. Shaver, 1970, Deltaic sedimentation, modern and ancient: Soc. Econ. Paleontol. and Mineral. Spec. Publ. 15.

Morisawa, M. E., 1959, Relation of quantitative geomorphology to stream flow in representative watersheds of the Appalachian Plateau province: Dept. Geol., Columbia Univ., Tech. Rept. 20.

Muller, G., 1966, The new Rhine Delta in Lake Constance: *In* M. L. Broussard, *ed.*, Deltas in their geologic framework: Houston, Texas; Houston Geol. Soc., pp. 107-124.

Murray, S. P., 1970, Settling velocities and vertical diffusion of particles in turbulent water: J. Geophys. Res., 75:1647-1654.

———, 1970, Bottom currents near the coast during Hurricane Camille: J. Geophys. Res., 75:4579-4582.

———, 1972, Observations on wind, tidal, and density-driven currents in the vicinity of the Mississippi River delta: *In* D. J. P. Swift, D. B. Duane, and O. H. Pilkey, *eds.*, Shelf sediment transport: Stroudsburg, Pa: Dowden, Hutchinson & Ross, pp. 126-142.

Nedeco, 1959, River studies and recommendations on improvement of Niger and Benue: Amsterdam, North Holland Publ. Co., 1000 pp.

Nelson, B. W., 1970, Hydrology, sediment dispersal, and Recent historical development of the Po River delta, Italy: *In* Morgan, J. P., *ed.*, Deltaic sedimentation, modern and ancient: Soc. Econ. Paleontol. and Mineral. Spec. Publ. 15, pp. 152-184.

Nissenbaum, A., B. J. Presley, and I. R. Kaplan, 1972, Early diagenesis in a reducing fjord, Saanich Inlet, B. C., I. Chemical and isotropic changes in major components of interstitial water: Geochim. et Cosmochim. Acta, 36:1007-1027.

Odem, W. I., 1953, Subaerial growth of the delta of the diverted Brazos River, Texas: Compass, 30(3): 172-178.

*Off, T. 1963, Rhythmic linear sand bodies caused by tidal currents: Am. Assoc. Petrol. Geologists Bull., 47:324-341.

Oomkens, E., 1967, Depositional sequences and sand distribution in a deltaic complex; a sedimentological investigation of the postglacial Rhone Delta complex: Geologie en Mijnb., 46:265-278.

———, 1974, Lithofacies relations in the Late Quaternary Niger Delta complex: Sedimentology, 21:195-221.

Ouellette, D. J., 1969, Sediment and water characteristics, South Pass, Mississippi River: Coastal Studies Inst., Louisiana State Univ., Baton Rouge, Bull. 3, pp. 29-53.

Peltier, L. C., 1950, The geographic cycle in periglacial regions as it is related to climatic morphology: Annals, Assoc. Am. Geogr., 40(3):214-236.

Peterson, J. A., and J. C. Osmond, *eds.*, 1961, Geometry of sandstone bodies: Am. Assoc. Petrol. Geologists, Tulsa, Okla., 240 pp.

Petrescu, I. G., 1948, Le delta maritime du Danube: son evolution physicogeographique et les problemes qui s'y posent: Univ. Jassy, Ann. Sci., 2nd sec. (Sc. Nat.), 31:254-303.

———, 1967, Geomorphology of the Danube Delta (trans.), Stuttgart: Archiv fur Hydrobiologie, Supplement, 30(4):322-339 (in German).

Pettijohn, F. J., and P. E. Potter, 1964, Atlas and glossary of primary sedimentary structures: New York, Springer-Verlag, 618 pp.

Pierson, W. J., G. Neumann, and R. W. James, 1967, Practical methods for observing and forecasting ocean waves by means of wave spectra and statistics: U. S. Naval Oceanogr. Office H. O. Publ. No. 603, 284 pp.

Potter, P. E., 1967, Sand bodies and sedimentary environments: a review: Am. Assoc. Petrol. Geologists Bull., 51:337-365.

———, and Pettijohn, F. J., 1963, Paleocurrents and basin analysis: New York, Springer-Verlag, 296 pp.

———, F. J. Pettijohn, and R. Siever, 1972, Sand and sandstone: New York, Springer-Verlag, 618 pp.

Potter, W. D., 1953, Rainfall and topographic factors that affect runoff: Trans. Am. Geophys. Union, 34:67-73.

Pryor, W. A., 1973, Permeability-porosity patterns and variations in some Holocene sand bodies: Am. Assoc. Petrol. Geologists Bull., 57:162-189.

Psuty, N. P., 1965, Beach-ridge development in Tabasco, Mexico: Annals, Assoc. Am. Geogr., 55:112-124.

Rainwater, E. H., 1975, Petroleum in deltaic sediments: *In* M. L. Broussard, *ed.*, Deltas, models for exploration, 2nd ed: Houston Texas; Houston Geol. Soc., pp. 3-11.

Razavet, C. D., 1956, Contribution a l'etude geologique et sedimentologique du delta du Rhone: Soc. Geol. France, Mem. 76.

Reimnitz, E., and N. F. Marshall, 1965, Effects of the Alaska earthquake and tsunami on recent deltaic sediments: J. Geophys. Res., 70:2363-2376.

Reineck, H. E., and I. B. Singh, 1973, Depositional sedimentary environments: New York, Springer-Verlag, 439 pp.

Reineck, H. E., and F. Wunderlich, 1968, Classification and origin of flaser and lenticular bedding: Sedimentology, 11:99-104.

Renz, O., R. Lakeman, and E. Van Der Meulen, 1955, Submarine sliding in western Venezuela: Am. Assoc. Petrol. Geologists Bull., 29:2053-2067.

Resio, D., L. Vincent, J. Fisher, B. Hayden, and R. Dolan, 1973. Classification of coastal environments; analysis across the coast barrier island interfaces: Dept. of Environmental Sci., Univ. of Virginia, Charlottesville, Tech. Rept. 5, 31 pp.

Richard, A. F., ed., 1967, Marine geotechnique: Proc., Internat. Res. Conf. on Marine Geotechnique, May 1-4, 1967, Urbana, University of Illinois Press.

Rigby, J. K., and W. K. Hamblin, 1972, Recognition of ancient sedimentary environments: Soc. Econ. Paleontol. and Mineral. Spec. Publ. 16, 340 pp.

Roberts, H. H., D. Cratsley, and T. Whelan, III, 1976, Stability of Mississippi Delta sediments as evaluated by analysis of structural features in sediment borings: Eighth Ann. Offshore Tech. Conf., Houston, Texas, May 3-5, 1976, pp. 9-28.

Rodolfo, K. S., 1975, The Irrawaddy Delta: Tertiary setting and modern offshore sedimentation: In M. L. Broussard, ed., Deltas, models for exploration, 2nd ed: Houston, Texas; Houston Geol. Soc., pp. 339-356.

Rouse, L. J., and J. M. Coleman, 1976, Circulation observations in the Louisiana Bight using LANDSAT imagery: Remote Sensing of Environment, 5:55-66.

*Russell, R. J., 1936, Physiography of the Lower Mississippi River delta: In Lower Mississippi Delta: Reports on the geology of Plaquemines and St. Bernard parishes. Louisiana Dept. of Conservation, Geol. Bull. 8, pp. 3-193.

―――, 1940, Quaternary history of Louisiana: Geol. Soc. Am. Bull., 51:1199-1234.

―――, 1942, Geomorphology of the Rhone Delta: Assoc. Am. Geogr. Annals., 32:149-254.

―――, 1954, Aluvial morphology of Anatolian rivers: Assoc. Am. Geogr. Annals., 44(4):363-391.

―――, 1958, Geological geomorphology: Geol. Soc. Am. Bull., 69(1):1-22.

―――, 1967. River and delta morphology: Coastal Studies Inst., Louisiana State Univ., Baton Rouge, Tech. Rept. 52, pp. 1-55.

―――, and R. D. Russell, 1939, Mississippi River delta sedimentation: In P. d. Trask, ed., Recent marine sediments: Tulsa, Oklahoma; Am. Assoc. Petrol. Geologists, pp. 153-177.

Samajilov, I. V., 1956, Die Flussmundungen: Gotha, Germany, Veb Hermann Haack, 647 pp. (trans. from Russian original Ust'ia Rek).

Saucier, R. T., 1963, Recent geomorphic history of the Pontchartrain Basin, Louisiana: Coastal Studies Inst., Louisiana State Univ., Baton Rouge, Coastal Studies Series No. 9, 114 pp.

Saxena, R. S., 1976, Modern Mississippi Delta—Depositional environments and processes: Am. Assoc. Petrol. Geologists—Soc. Econ. Paleontol. and Mineral. field trip guidebook. New Orleans, Louisiana, May 23-26, 1975, 125 pp.

―――, ed., 1976, Sedimentary environments and hydrocarbons: Am. Assoc. Petrol. Geologists short course, New Orleans, Louisiana, May 23, 1976, 217 pp.

―――, and J. C. Ferm, 1976, Lower deltaic sand bodies—exploration models and recognition criteria: In R. S. Saxena, ed., Sedimentary environments and hydrocarbons: Am. Assoc. Petrol. Geologists short course, New Orleans, Louisiana, May 23, 1976, pp. 22-59.

Scarpace, F. L., and T. Green, III, 1973, Dynamic surface temperature structure of thermal plumes: Water Resources Res., 9:138-153.

Schumm, S.A., 1954, Evolution of drainage systems and slopes in badlands at Perth Amboy, New Jersey: Dept. of Geol., Columbia Univ., New York, Tech. Rept. 8, ONR Contract N6ONR271-30.

―――, 1960, The shape of alluvial channels in relation to sediment type: U.S. Geol. Surv. Prof. Paper 351-B, pp. 17-30.

―――, 1963, Sinuosity of alluvial rivers on the Great Plains: Geol. Soc. Am. Bull., 79:1089-1100.

Scruton, P. C., 1955, Sediments of the eastern Mississippi Delta: In J. L. Hough and H. W. Menard, eds., Finding ancient shorelines, Soc. Econ. Paleontol. and Mineral. Spec. Publ. 3, pp. 21-51.

―――, 1956, Oceanography of Mississippi Delta sedimentary environments: Am. Assoc. Petrol. Geologists Bull., 40:2864-2952.

*―――, 1960, Delta building and the deltaic sequence: In F. P. Shepard et al., eds., Recent sediments, northwest Gulf of Mexico: Tulsa, Oklahoma, Am. Assoc. Petrol. Geologists. pp. 82-102.

―――, and D. G. Moore, 1953, Distribution of surface turbidity off the Mississippi Delta: Am. Assoc. Petrol. Geologists Bull., 37:1067-1074.

Selley, R. C., 1970, Ancient sedimentary environmments. London, Chapman & Hall, 237 pp.

Shepard, F. P., 1937, Revised classification of marine shorelines: J. Geol., 45:602-624.

―――, 1952, Revised nomenclature for depositional coastal features: Am. Assoc. Petrol. Geologists Bull., 36:1902-1912.

―――, 1955, Sediment zones bordering the barrier islands of central Texas coast: In Finding ancient shorelines, pp. 78-97, Soc. Econ. Paleontol. and Mineral. Spec. Publ. 3.

*―――, 1955, Delta front valleys bordering the Mississippi distributaries: Geol. Soc. Am. Bull., 66:1489-1498.

―――, 1956, Late Pleistocene and Recent history of the central Texas coast: J. Geol., 64:56-69.

―――, 1956, Marginal sediments of the Mississippi Delta: Am. Assoc. Petrol. Geologists Bull., 40:2537-2623.

———, 1960, Mississippi Delta: Marginal environments, sediments and growth: *In* Recent sediments, northwestern Gulf of Mexico. Am. Assoc. Petrol. Geologists.

———, 1973, Submarine geology, 3rd eds., New York, Harper and Row, 517 pp.

*———, 1973, Sea floor off Magdalena Delta and Santa Marta area, Colombia: Geol. Soc. Am. Bull., 84:1955-1972.

———, R. F. Dill, and B. C. Heezen, 1968, Diapiric intrusions in foreset slope sediments off Magdalena Delta, Colombia: Am. Assoc. Petrol. Geologists Bull., 42:2197-2207.

———, and K. O. Emery, 1973, Congo submarine canyon and fan valley: Am. Assoc. Petrol. Geologists Bull., 57:1679-1691.

———, and D. G. Moore, 1955, Sediment zones bordering the barrier islands of central Texas Coast: *In* J. L. Hough and H. W. Menard, *eds.*, Finding ancient shorelines, a symposium, Soc. Econ. Paleontol. and Mineral. Spec. Publ. 3, pp. 78-98.

——— And D. G. Moore, 1955, Central Texas coast sedimentation; characteristics of sedimentary environments, recent history and diagenesis. Am. Assoc, Petrol. Geologists Bull., 39:1563-1593

———, F. B. Phleger, and Tj. H. van Andel, *eds.*, 1960, recent sediments, northwest Gulf of Mexico: Am. Assoc. Petrol. Geologists, 394 pp.

———, and H. R. Wanless, 1971, Our changing coastlines: New York, McGraw-Hill, 579 pp.

———, and R. Young, 1961, Distinguishing between beach and dune sands: J. Sediment. Petrol., 31:196-214.

Shirley, M. L., *ed.*, 1966, Deltas. Houston Geol. Soc: Houston, Texas, 251 pp.

Silvester, R., and C. LaCruz, 1970, Pattern forming processes in deltas: J. Waterways and Harbours Div., Am. Soc. Civil Engr., 96(WW2):201-217.

Singh, H., 1974, The effects of waves on ocean sediments: Dames and Moore Eng. Bull. 44, pp. 11-21.

Smith, A. E., Jr., 1966, Modern deltas: comparison maps: *In* M. L. Shirley, *ed.*, Deltas in their geologic framework: Houston, Texas; Houston Geol. Soc., p. 233.

———, comp., and M. L. Broussard, *ed.*, 1971, Deltas of the world: modern and ancient: bibliography: Houston Geol. Soc., Delta Study Group, 42 pp.

Snead, R. E., 1964, Active mud volcanoes of Baluchistan, West Pakistan: Geogr. Review. LIV(4):546-560.

Sonu, C. J., S. P. Murray, S. A. Hsu, J. N. Suhayda, and E. Waddell, 1973, Sea breeze and coastal processes: EOS Trans. Am. Geophys. Union, 54:820-833.

Sorbey, H. C., 1859, On the structure produced by the currents present during the deposition of stratified rocks: The Geologist, 2:137-147.

Spearing, D. R., 1975, Summary sheets of sedimentary deposits: Geol. Soc. Am. Map and Chart Series MC-8.

Sprigg, R. C., 1950, The geology of the South-East Province, South Australia, with special reference to Quaternary coast-line migrations and modern beach developments: Bull. Geol. Surv. South Australia, Adelaide, 29.

Stanley, D. J., and D. J. P. Swift, *eds.*, 1974, The new concepts of continental margin sedimentation, II, Sediment transport and its application to environmental management: Lecutre notes, AGI Short Course, No. 15-17, 1974, Key Biscayne, Florida, Parts 1 and 2, 1155 pp.

Stephens, D. G., D. S. Van Niewenhuise, P. Mullins, C. Lee, and W. H. Kanes, 1975, Destructive phase of deltaic development: North Santee River Delta: J. Sediment. Petrol., 46:132-144.

Steinmetz, R., 1967, Depositional history primary sedimentary structures, cross bed dips, and grain size of an Arkansas River point bar at Wekiwa, Oklahoma: Amoco Prod. Co., Rept. F67-G-3.

———, 1972, Sedimentation of an Arkansas River sand bar in Oklahoma: a cautionary note on dipmeter interpretation: Shale Shaker, 23(2):32-38.

*Sterling, G. H., and E. E. Strohbeck, 1973, The failure of the South Pass 70 "B" platform in Hurricane Camille: Preprints, Offshore Tech. Conf., Houston, Texas, April 1973.

Stolzenback, K. D., and D. R. F. Harleman, 1971, An analytical and experimental investigation of surface discharges of heated water: Massachusetts Inst. of Tech., Ralph M. Parsons Lab Water Resources and Hydrodynamics, Dept. of Civil Eng., Rept. 135, 212 pp.

———, and D. R. F. Harleman, 1973, Three-dimensional heated surface jets: Water Resources Res., 9:129-137.

Stommel, H., and G. H. Farmer, 1952, Abrupt change in width in two-layer open channel flow: J. Marine Res., 11:205-214.

Straaten, L. M. J. U. van, 1954, Composition and structure of Recent marine sediments in the Netherlands: Leidse Geol., Mededl. 19, pp. 1-1110.

———, 1957, Recent sandstones on the coast of the Netherlands and of the Rhone Delta: Geol. en Mijnb., 19:196-213.

———, 1957, The excavation at Velsen, general introduction: *In* The excavation at Velsen. Verhandl. Koninkl. Ned. Geol. Mijnb. Gen. Ser. 17:93-99.

———, 1959, Littoral and submarine morphology of the Rhone Delta: Proc., 2nd Coastal Geogr. Conf., Baton Rouge (Nat. Acad. Sci.—Nat. Res. Council), pp. 233-264.

———, 1959, Minor structures of some recent littoral and neritic sediments: Geol. en Mijnb., 21:197-216.

———, 1960, Some recent advances in the study of deltaic sedimentation: Liverpool, Manchester, Geol. J., 2:411-442.

———, *ed.*, 1964, Deltaic and shallow marine deposits: Developments in sedimentology. Amsterdam, Elsevier, v. 1, 464 pp.

———, 1970, Holocene and Late Pleistocene sedimentation in the Adriatic Sea: Geol. Rundschau 60, pp. 106-131.

Strahler, A. N., 1952, Hypsometric (area altitude)

analysis of erosional topography: Geol. Soc. Am. Bull., 63:1117-1142.

———, 1957, Quantitative analysis of watershed geomorphology: Am. Geophys. Union Trans. 38(6):913-920.

———, 1958, Dimensional analysis applied to fluvially eroded landforms: Geol. Soc. Am. Bull., 69:279-300.

*Suhayda, J. N., T. Whelan, III, J. M. Coleman, J. S. Booth, and L. E. Garrison, 1976, Marine sediment instability: interaction of hydrodynamic forces and bottom sediments: Eighth Ann. Offshore Tech. Conf., Houston, Texas May 3-6, 1976, Paper OTC 2426, pp. 29-40.

Sundborg, A., 1956, The Kiver Klaralven: a study of fluvial processes: Geogr. Ann., Stockholm, 38:127-316.

Swan, S. B. St. C., 1968, Coastal classification with reference to the east coast of Malaya: Zeit. fur Geomorphologie, 7:114-132.

Sykes, G. G., 1937, The Colorado Delta: Carnegie Inst., Washington, D. C., Publ. 460, 193 pp.

Takai, Y., and T. Kamura, 1966, The mechanism of reduction in water-logged paddy soil: Folia Microbiol. Prague, 11:304-313.

Takano, K., 1954, On the velocity distribution off the mouth of a river, I: J. Oceanogr. Soc. Japan, 10:60-64.

———, 1954, On the salinity and velocity distributions off the mouth of a river II: J. Oceanogr. Soc. Japan, 10:92-98.

———, 1955, A complementary note on the diffusion of the seaward flow off the mouth of a river: J. Oceanogr. Soc. Japan, 11:1-3.

Thom, B. G., L. D. Wright, and J. M. Coleman, 1975, Mangrove ecology and deltaic-estuarine geomorphology: Cambridge Gulf-Ord River, Western Australia: J. Ecology, 63: 203-232.

Tricart, J., 1955, Geomorphologic map of Senegal Delta: Acad. Geol. Francaise Bull., No. 251, pp. 98-117.

———, 1955, Aspects sedimentologiques du delta du Senegal: Geol. Rundschau, 43(2): 384-397.

———, 1956, Aspects geomorphologiques du delta du Senegal: Revue de Geomorphol. Dynamique, 7(5-6):65-86. Also, Assoc. Geogr. Francais, Bull. Nos. 251-252, pp. 98-117.

———, 1955, Aspects sedimentologiques du delta du Senegal: Geol. Rundschau, 43(2): 384-397.

———, 1956, Aspects geomorphologiques du delta du Senegal: Revue de Geomorphol. Dynamique, 7(5-6):65-86. Also, Assoc. Geogr. Francais, Bull. Nos. 251-252, pp. 98-117.

———, 1959, Presentation of a sheet of the 1:50,000 geomorphological map of the Senegal Delta: Revue de Geomorphol. Dynamique, 10:106-116.

———, and A. Cailleaux, 1965, Introduction a la geomorphologie climatique (Introduction to climatic geomorphology): Paris, Societe de'Edition d'Enseignement Superieur, 306 pp.

Trowbridge, A. C., 1930, Building of the Mississippi Delta: Am. Assoc. Petrol. Geologists Bull., 14:85-107.

*Tubman, M. W., and J. N. Suhayda, 1976, Wave action and bottom movements in fine sediments: Proc., 15th Conf. on Coastal Engr., ASCE, July 11-17, 1976, Honolulu, Hawaii, pp. 1168-1175.

Van Lopik, J. R., 1955, Recent geology and geomorphic history of central coastal Louisiana: Coastal Studies Inst., Louisiana State Univ., Baton Rouge, Tech. Rept. 7, 89 pp.

Visher, G. S., S. B. Ekebafe, and J. Rennison, The Coffeyville format (Pennsylvanian) of northeastern Oklahoma, a model for an epeiric sea delta: *In* M. L. Broussard, *ed.*, Deltas, models for exploration, 2nd ed: Houston, Texas; Houston Geol. Soc., pp. 381-397.

Volker, A., 1966, Tentative classification and comparison with deltas of other climatic regions: *In* Scientific problems of the humid tropical zone deltas and their implications: UNESCO, Proc. Dacca Symp., pp. 399-408.

Waldrop, W. R., 1972, Three-dimensional flow and sediment transport at river mouths: Unpub. Ph.D. dissertation, Louisiana State Univ., Baton Rouge, 233 pp.

———, 1973, Preliminary river-mouth flow model: Coastal Studies Inst., Louisiana State Univ., Bull. 7, pp. 67-92.

Walker, H. J., 1961, The Colville River delta: First Nat. Coastal and Shallow Water Res. Conf., pp. 472-474.

*Watson, J. G., 1928, Mangrove forests of the Malay Peninsula: Singapore, Malayan Forest Record 6, 275 pp.

Welder, F. A., 1959, Processes of deltaic sedimentation in the Lower Mississippi River: Coastal Studies Inst., Louisiana State Univ., Baton Rouge, Tech. Rept. 12, 90 pp.

*Wells, T., D. Prior, J. M. Coleman, 1980, Slow-slides in extremely low angle tidal flat muds, Northeast South America, Geology, v. 8, pp. 272-275.

West, R. C., N. P. Psuty, and B. G. Thom, 1969, The Tabasco lowlands of southeastern Mexico: Coastal Studies Inst., Louisiana State Univ., Coastal Studies Series No. 27, 193 pp.

*Whelan, T., III, J. M. Coleman, H. H. Roberts, and J. N. Suhayda, 1976, Occurrence of methane in Recent deltaic sediments and its effect on soil stability: Bull. International Assoc. of Engr. Geology, 14 pp. 55-64.

*———, J. M. Coleman, J. N. Suhayda, and L. E. Garrison, 1975, The geochemistry of Recent Mississippi River delta sediments: gas concentration and sediment stability: Preprints, 7th Ann. Offshore Tech. Conf., Houston, Texas, May 5-8, 1975, pp. 71-84.

Wiseman, Wm. J., Jr., J. M. Coleman, A. Gregory, S. A. Hsu, A. D. Short, J. N. Suhayda, C. D. Walters, Jr., and L. D.

Wright, 1973, Alaskan Arctic coastal processes and morphology: Coastal Studies Inst., Louisiana State Univ., Tech. Rept. 149, 171 pp.

Wishart, D., 1969, FORTRAN II programs for 8 methods of cluster analysis (CLUSTAN I): Kansas Geol. Surv., Computer Contr. 38, 111 pp.

Wright, L. D., 1970, Circulation, effluent diffusion, and sediment transport, mouth of South Pass, Mississippi River delta: Coastal Studies Inst., Louisiana State Univ., Baton Rouge, Tech. Rept. 84, 56 pp.

*_____, and J. M. Coleman, 1971, Effluent expansion and interfacial mixing in the presence of a salt wedge, Mississippi River delta: J. Geophys. Res., 76:8649-8661.

_____, and J. M. Coleman, 1972, River delta morphology, wave climate and the role of the subaqueous profile: Science, 176:282-284.

*_____, and J. M. Coleman, 1973, Variation in morphology of major river deltas as functions of ocean waves and river discharge regimes: Am. Assoc. Petrol. Geologists Bull., 47:370-398.

*_____, and J. M. Coleman, 1974, Mississippi River mouth processes: effluent dynamics and morphologic development: J. Geol., 82:751-778.

*_____, J. M. Coleman, and M. W. Erickson, 1974, Analysis of major river systems and their deltas: morphologic and process comparisons: Coastal Studies Inst., Louisiana State Univ., Tech. Rept. 156, 114 pp.

*_____, J. M. Coleman, and J. N. Suhayda, 1973, Periodicities in interfacial mixing: Coastal Studies Inst., Louisiana State Univ., Bull. 7, pp. 127-135.

_____, J. M. Coleman, and B. G. Thom, 1972, Emerged tidal flats in the Ord River Estuary, Western Australia: Search, 3(9): 339-341.

_____, J. M. Coleman, and B. G. Thom, 1973, Processes of channel development in a high-tide-range environment: Cambridge Gulf-Ord River delta, Western Australia: J. Geol., 81:15-41.

*_____, J. M. Coleman, and B. G. Thom, 1975, Sediment transport and deposition in a macrotidal river channel: Ord River, western Australia: *In* Estuarine research, v. II, Geology and engineering: New York, Academic Press, pp. 309-321.

_____, F. J. Swaye, and J. M. Coleman, 1970, Effects of Hurricane Camille on the landscape of the Breton-Chandeleur Island chain and the eastern portion of the Lower Mississippi Delta: Coastal Studies Inst., Louisiana State Univ., Baton Rouge, Bull. 4, pp. 13-34.

Zenkovitch, V. P., 1967, Processes of coastal development: New York, John Wiley, 738 pp.

*References cited in text.